Jochen Lohmiller

Investigation of deformation mechanisms in nanocrystalline metals and alloys by in situ synchrotron X-ray diffraction

**Schriftenreihe
des Instituts für Angewandte Materialien**

Band 13

Karlsruher Institut für Technologie (KIT)
Institut für Angewandte Materialien (IAM)

Eine Übersicht über alle bisher in dieser Schriftenreihe erschienenen Bände
finden Sie am Ende des Buches.

Investigation of deformation mechanisms in nanocrystalline metals and alloys by in situ synchrotron X-ray diffraction

by
Jochen Lohmiller

Dissertation, Karlsruher Institut für Technologie (KIT)
Fakultät für Maschinenbau, 2012
Tag der mündlichen Prüfung: 11. Dezember 2012

Impressum

Karlsruher Institut für Technologie (KIT)
KIT Scientific Publishing
Straße am Forum 2
D-76131 Karlsruhe
www.ksp.kit.edu

KIT – Universität des Landes Baden-Württemberg und
nationales Forschungszentrum in der Helmholtz-Gemeinschaft

KIT Scientific Publishing 2013
Print on Demand

ISSN 2192-9963
ISBN 978-3-86644-962-6

Investigation of deformation mechanisms in nanocrystalline metals and alloys by in situ synchrotron X-ray diffraction

Zur Erlangung des akademischen Grades

Doktor der Ingenieurwissenschaften

der Fakultät für Maschinenbau
Karlsruher Institut für Technologie (KIT)

genehmigte

Dissertation

von

DIPL.-ING. JOCHEN ANDREAS LOHMILLER

aus Herrenberg

Tag der mündlichen Prüfung: 11.12.2012
Hauptreferent: Prof. Dr. Oliver Kraft
Korreferent: Prof. Dr. Ralph Spolenak

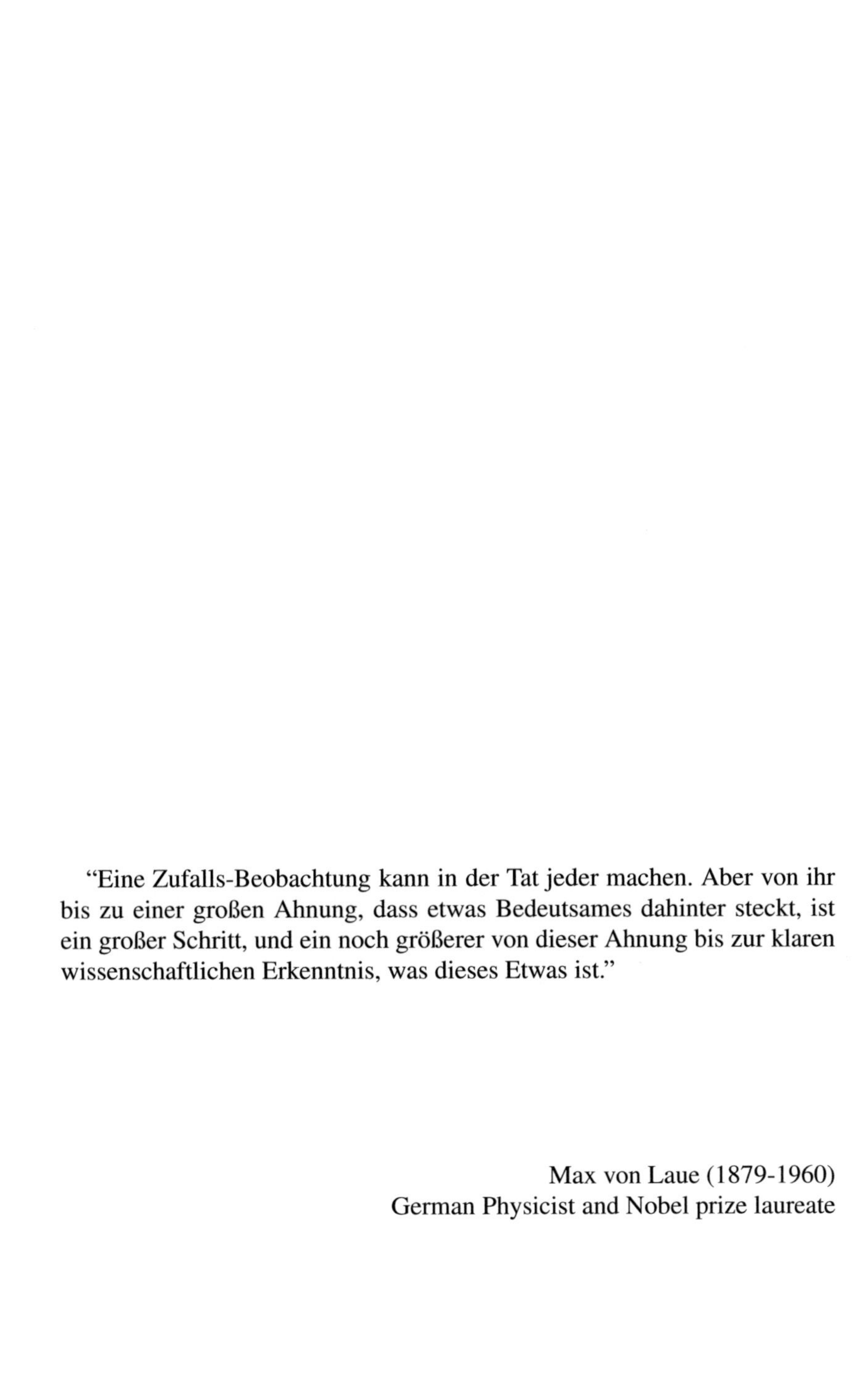

"Eine Zufalls-Beobachtung kann in der Tat jeder machen. Aber von ihr bis zu einer großen Ahnung, dass etwas Bedeutsames dahinter steckt, ist ein großer Schritt, und ein noch größerer von dieser Ahnung bis zur klaren wissenschaftlichen Erkenntnis, was dieses Etwas ist."

Max von Laue (1879-1960)
German Physicist and Nobel prize laureate

Kurzzusammenfassung

Voraussetzung für die Ausnutzung der vielversprechenden Eigenschaften von nanokristallinen Metallen (Korngröße $D < 100\,\text{nm}$) ist das Verständnis der zugrunde liegenden Verformungsmechanismen. Durch die Verringerung der Korngröße wird die mechanische Festigkeit gesteigert. Dadurch wird konventionelle Versetzungsplastizität behindert und andere Verformungsmechanismen können dominant werden. Demzufolge wurden verschiedenene Mechanismen vorgeschlagen: Andere intragranulare Mechanismen, wie das Gleiten von Partialversetzungen oder Zwillingsbildung, oder intergranulare Mechanismen, wie Korngrenzenmigration oder Korngrenzengleiten. Es ist anzunehmen, dass insbesondere für Korngrößen kleiner als $30\,\text{nm}$ intergranulare Mechanismen eine bedeutende Rolle einnehmen.

Das Ziel dieser Dissertation war die systematische Untersuchung von Verformungsmechanismen verschiedener nanokristalliner Metalle und Legierungen in diesem Korngrößenbereich mit Hilfe von weiterentwickelten Synchrotron-basierten *in situ* mechanischen Prüfmethoden. Getestet wurden sowohl Massivproben aus Ni ($D \approx 30\,\text{nm}$) und verschiedenen PdAu-Legierungen ($D \approx 10\,\text{-}15\,\text{nm}$) unter Druck- und Scher-Druckbelastung als auch Polyimid unterstütze dünne Schichten aus reinem Pd und verschiedenen PdAu-Legierungen ($D \approx 20\,\text{-}\,30\,\text{nm}$) unter Zugbelastung. Die hochentwickelte Auswerteroutine von Röntgenbeugungsprofilen ermöglicht es sowohl aufeinanderfolgende, gleichzeitig auftretende, als auch reversible Verformungsmechanismen identifizieren, separieren und den entsprechenden Dehnungsbereichen zuordnen zu können.

Charakteristische Merkmale an mehreren unabhängigen Beugungsparametern wurden identifiziert. Beispielsweise lässt sich Versetzungsplastizität durch die Umkehr (hkl) unabhängiger asymmetrischer Profilformen und der Bildung von Verformungstexturen nachweisen. Änderungen der Korngröße und der Kornform wurden mittels Profilbreitenanalyse in Kombination mit Flächendetektordaten verfolgt. Isotropes Kornwachstum belegt das Auftreten von spannungsinduzierter

Korngrenzenmigration, wohingegen die Entwicklung einer elliptischen Kornform wiederum auf Versetzungsplastizität hindeutet.

Alle Proben zeigten eine Koexistenz und ein Aufeinanderfolgen von verschiedenen Verformungsmechanismen. Charakteristisch für nanokristalline Metalle sind die inhärent hohen, elastischen Gitterdehnungen aufgrund der hohen Festigkeiten und die von Beginn an resultierenden Akkommodierungsmechanismen an Korngrenzen und/oder Tripellinien. Des Weiteren wurde mechanisch getriebenes Kornwachstum in allen getesteten Proben nachgewiesen. Die Ausprägung und Anteile der Korngrenzenmigration variierten von Probentyp zu Probentyp stark, was auf die unterschiedliche Reinheit der Herstellungsmethoden zurückgeführt wurde. Versetzungsplastizität in Ni und Pd trägt deutlich zur Gesamtverformung bei, jedoch ist die Versetzungsaktivität in den PdAu-Massivproben aufgrund der äußerst geringen Korngröße sehr stark eingeschränkt. Verglichen mit Ni zeigte Pd weniger Versetzungsplastizität und mehr Korngrenzenmigration. Unter scherdominierter Verformung erhöhten sich die Anteile versetzungsbasierter Mechanismen. Legierungseffekte auf die Entstehung von Verformungstexturen und spannungsgetriebenem Kornwachstum wurden identifiziert. Stärkere Zunahmen wurden bei höheren Goldgehältern in den PdAu-Legierungen gemessen. Legierungen mit geringem Goldgehalt zeigten andererseits spröderes Materialverhalten mit Rissbildung.

Abstract

A prerequisite for the exploitation of the promising properties of nanocrystalline metals (grain size $D < 100\,\text{nm}$) is the understanding of the underlying mechanisms. By reducing the grain size, an increase of the mechanical strength is achieved. In doing so, conventional dislocation plasticity is inhibited and other deformation mechanisms may become dominant. Therefore, various mechanisms have been proposed: Other intragranular mechanisms, such as motion of partial dislocations or twinning, or intergranular mechanisms, such as grain boundary migration or grain boundary shear and slip. Especially for grain sizes smaller than 30 nm it is expected that intergranular deformation mechanisms play a prominent role.

The goal of this thesis was to systematically investigate the deformation mechanisms of different nanocrystalline metals and alloys in this grain size range using advanced synchrotron-based *in situ* mechanical testing. The tested materials are bulk samples of Ni ($D \approx 30\,\text{nm}$) and different PdAu alloys ($D \approx 10$ -$15\,\text{nm}$) under compressive and shear-compressive loading, as well as, polyimide-supported thin films consisting of pure Pd and various PdAu alloys ($D \approx 20$ - $30\,\text{nm}$) under tensile loading. Sophisticated XRD peak profile analysis allows for the identification of succeeding, coexisting, as well as reversible deformation mechanisms and furthermore their separation and classification to specific strain regimes. Characteristic features of several independent peak parameters were found: E.g. the reversal of (hkl)-independent asymmetric peak profiles and the formation of deformation textures are evidence of dislocation-based plasticity. In-plane changes of grain size and grain shape were followed by peak broadening analysis in combination with diffraction data obtained by area detectors. Isotropic grain growth reveals the occurrence of stress-driven grain boundary migration, whereas an emerging elliptic grain shape again points to dislocation plasticity.

All samples exhibit a coexistence and succession of different deformation mechanisms. Characteristic for NC metals are the inherently large lattice strains, given by the high strength, and the - from the beginning - resulting accommodation processes in grain boundaries and/or triple lines. Moreover, deformation-induced grain growth was identified in all samples. The occurrence and contribution of grain boundary migration was strongly varying from sample type to sample type, which was attributed to the differently "clean" preparation routes. Dislocation-based plasticity contributes considerably to overall deformation for Ni and Pd, however, it is highly aggravated for the PdAu alloys due to the very small grain size. Compared to Ni, Pd demonstrates less dislocation plasticity but enhanced grain boundary migration. Shear-dominated deformation promotes dislocation-based plasticity and increases its relative contribution. Alloying effects were identified for the formation of deformation textures and stress-induced grain growth. More pronounced increases were measured for higher Au contents in PdAu alloys. On the other hand, alloys with low Au content showed more brittle behavior with crack formation.

Danksagung

Die vorliegende Arbeit wurde in der Zeit von November 2009 bis Oktober 2012 am Institut für Angewandte Materialien (IAM) am Karlsruher Institut für Technologie (KIT) angefertigt.

Herrn Prof. Dr. Oliver Kraft danke ich für die Annahme als Doktorand an seinem Institut, die wertvollen und hilfreichen fachlichen Kommentare und Diskussionen, sowie die Möglichkeit an vielen internationalen und nationalen Tagungen teilnehmen zu können.

Bei Herrn Prof. Dr. Ralph Spolenak bedanke ich mich für die Übernahme des Mitberichts und sein stetes Interesse an meiner Arbeit. Für seine Hilfestellung und seine Unterstützung, auch vor meiner Doktorarbeit, bin ich ihm sehr dankbar.

Herrn Dr. Patric Gruber bin ich dankbar für die fachliche Betreuung, die gute Zusammenarbeit, seine Diskussionbereitschaft und die Unterstützung bei den gemeinsamen Strahlzeiten. Dadurch konnte ich sehr viel lernen und meine Arbeit hat entscheidend profitiert. Außerdem genoß ich große wissenschafltiche Freiheit um meine eigenen Ideen zu verfolgen.

Herrn Rudolf Baumbusch danke ich für den ständigen wissenschaftlichen Austausch und die grundlegenden Arbeiten an der Auswerteroutine.

Meinen Kooperationspartnern danke ich für ihre individuelle Unterstützung: Bei Prof. Dr. Rainer Birringer, Manuel Grewer und Christian Braun bedanke ich mich für viele hilfreiche Diskussionen. Aaron Kobler und Dr. Christian Kübel danke ich für die Untersuchungen mittels ACOM/TEM. Für die Herstellung der verschiedenen Proben möchte ich mich bei Kerstin Schüler (Ni), Anna Castrup (Pd Schichten) und der Gruppe von Rainer Birringer (PdAu Massivproben) bedanken.

Dem Personal der einzelnen Strahlrohre (ID15A (ESRF, Grenoble), MS04 (SLS, Villigen) und MPI-MF (ANKA, Karlsruhe)) bin ich für die exzellente Unterstützung dankbar, besonderer Dank gilt Dr. Veijo Honkimäki.

Bei den Mitarbeitern des IAM-WBM bedanke ich mich für die angenehme Arbeitsatmosphäre und die zahlreichen Freizeitaktivitäten. Ich glaube man kann sich kaum bessere Kollegen vorstellen.

Meiner Familie, insbesondere meinen Eltern und meiner Freundin Michi, bin ich zu tiefstem Dank verpflichtet. Sie tragen einen nicht zu vernachlässigenden Anteil am Erfolg meiner Arbeit durch ihre unmessbare Unterstützung, Motivation und ihr Vertrauen in mich.

Contents

1. Introduction

Nanotechnology is a widespread, interdisciplinary field involving physics, chemistry, biology, and computer sciences. Molecular nanotechnology, nanoelectronics, nanobionics, molecular self-assembly, or nanomaterials are only a few keywords. Many of the most recent Nobel Prizes in Chemistry and Physics were related to nanotechnology (or even to nanomaterials), thus underlining the scientific and technological relevance (e.g. Geim/Novoselov, "two-dimensional graphene", Nobel Prize in Physics 2010 [Geim and Novoselov, 2007]).

Pioneered by Herbert Gleiter [Gleiter, 1984; Birringer et al., 1984], nanotechnology found its way into materials science in the early 1980's and the term "nanocrystalline" (NC) was created. Upon solidification, metal atoms usually arrange in a periodic structure with defined and repetitive relationships of neighboring atoms, in other words they crystallize. A large quantity of metals crystallizes in the face-centered cubic (FCC) crystal structure. FCC metals are interesting for many technical applications due to their good formability. Dislocations, which are the main carrier of plastic deformation can operate efficiently in the FCC lattice, since it is closest packed and usually offers a sufficient number of primary glide systems. During solidification usually numerous nuclei exist, and consequently a polycrystalline structure is obtained. The contiguous crystals / grains are separated by grain boundaries (GBs), which disrupt the periodic ordering. The disordered lattice structure within GBs and the change of grain orientation between adjacent grains impedes the unhindered motion of dislocations. The crystal size / grain size (D) can be systematically regulated via fabrication routes, which are nowadays quite divers. Reducing the grain size, and hence confining the range for dislocations to operate, yields a strengthening effect. It is proposed, that below a certain grain size, dislocations can no longer operate and other **deformation mechanisms** take over. Nowadays, a frequently used definition for **NC metals** is a grain size of less than 100 nm. In this regime, the transition of deformation mechanisms is expected.

Aside from grain refinement, solute atoms can also influence the mechanical performance of polycrystalline structures, yielding a strengthening effect. Traditionally **alloying** plays an important role in classical metallurgy as, for instance, manifested by the superiority of steels compared to pure iron.

An ideal method to probe a crystalline material nondestructively is **X-ray diffraction (XRD)**. Exactly 100 years after Max von Laue discovered the diffraction of X-rays by crystals in 1912, modern materials science can hardly be imagined without XRD and it has become an indispensable method for materials characterization. The periodicity of illuminated atoms is causing constructive and destructive interference, yielding a diffraction pattern that consists of intensity profiles. Changes of the periodic arrangement of the atoms, e.g. due to deformation of the structure, can cause a modified diffraction pattern, in which the positions and shapes of the intensity profiles change. *In situ* experiments allow to follow these modifications "live" during deformation.

In this thesis, NC FCC metals and alloys, mechanical testing, and *in situ* XRD are combined, to aim at identifying physical deformation mechanisms during mechanical testing of NC metals and alloys by sophisticated XRD peak shape analysis.

1.1. Motivation

Reducing the grain size of a polycrystal may offer the unique combination of increasing strength and toughness. NC materials are more and more used for structural applications, as well as for magnetic, electrical, and chemical applications. In order to exploit the full potential of NC materials for high performance tasks, the mechanical behavior must be understood. This necessitates the identification of the dominant deformation mechanisms and the succession of the related underlying microscopic processes, particularly in the lower NC regime ($D \approx 10 - 30\,$nm), where ordinary dislocation plasticity should be restricted and - at least partially - replaced by alternative mechanisms.

In situ experiments are necessary, since it was proposed that some deformation mechanisms do not leave a footprint after deformation [Budrovic et al., 2004]. *In situ* XRD, using 3rd generation synchrotron sources with high brilliance in combination with fast detectors, samples over a large volume and provides insight on the average material behavior and microstructural evolution [Van Swygenhoven et al., 2006]. A routine for peak shape analysis, going beyond the state-of-the-art incorporating asymmetric peak profiles, is employed to identify succeeding and coexisting deformation mechanisms during loading based on the evolution of peak position, broadening, asymmetry, and intensity.

Since NC structures are fabricated by a spectrum of different methods, which all have their individual character and peculiarities, results and interpretations of different studies are often contradictory. Therefore, special attention is paid to this issue by carefully characterizing the microstructures of the NC metals and alloys tested in this thesis. Results from *in situ* XRD, as the main experimental technique, are approved by complementary microscopic analyses in the Scanning Electron Microscope (SEM) and the Transmission Electron Microscope (TEM). Based on the comprehensive insights, microstructure-property relationships can be established.

1.2. Outline

The thesis is organized in the following way. In chapter 2, the most important literature is reviewed, including size effects due to dimensional and microstructural constraints as well as a detailed discussion on deformation mechanisms which have been proposed to explain the unique mechanical properties of NC metals. Chapter 3 covers all methodological aspects: Explaining the utilized experimental setups, the elaborate data analysis, and sample preparation. The investigated materials can be classified into bulk samples (tested under compressive and shear-compressive loading) and polyimide-supported thin films (tested under tensile loading). Chapter 4 examines the deformation behavior of bulk NC metals and alloys, with special focus on succeeding deformation mechanisms, and thereto related effects of different loading conditions, alloying effects, and grain size effects. Similarly to chapter 4, the tensile deformation behavior of NC thin films is researched in chapter 5. Likewise, the succession and coexistence of different mechanisms is demonstrated, and alloying effects on the initial microstructure and the mechanical response are discussed. In chapter 6, the key findings of the individual chapters are compared and cross-linked in order to derive general trends for the mechanical behavior of NC metals, but also to identify material-specific behavior. Finally, the thesis is summarized in chapter 7.

Parts of chapter 4 (basically the parts concerning the pure compression experiments of Ni) are in the process of publishing (see Ref. [Lohmiller et al., 2012b]). Parts of chapter 5 inspecting the behavior of pure Pd with a line detector will be published in Ref. [Lohmiller et al., 2013].

2. Literature Review

2.1. Size Effects

The examination of size effects has a long standing history in academic materials science and engineering. Based on the early work by Hall [Hall, 1951] and Petch [Petch, 1953], the following scaling law became apparent and can be frequently found in literature

$$\sigma_y = \sigma_0 + kd^n \tag{2.1}$$

where σ_y is the yield strength, σ_0 is the bulk strength, d is the characteristic length scale, and k and n are constants. Generally, with decreasing the characteristic length scale strength increases ($n < 0$). Traditionally, for polycrystalline materials, d is the grain size and n is in the range of -0.5, referred to as the classical Hall-Petch (HP) relation. The bulk deformation behavior is governed by the classical deformation mechanisms of dislocation glide, pile-up and interaction. Limiting the dislocations operation range aggravates pile-up and interaction of dislocations and thereby yields a strengthening effect, since the activation and/or nucleation of further dislocations requires an increasing value of stress. In a polycrystalline structure, grain boundaries (GBs) are the barriers for ordinary dislocation activity.

In recent years it was shown that this scaling law holds true for gradually decreasing characteristic length scale d, even when approaching the lower micrometer range. Often, the restraints lead to reduced ductility for smaller d, as a consequence of limited strain hardening since pile-ups and dislocation-dislocation reactions become scarce.

If the reduction of d is continued to the sub-micrometer range ($d <$ 1 µm), the strong size dependence attenuates, as a result of different deformation behavior. Assuming a common dislocation density of $\rho = 10^{12}$-10^{13} m^{-2}, the mean distance of dislocations is 1 µm or less and hence in the order of the characteristic length scale. As a result, deformation is carried by individual dislocations and mechanisms such as multiplication or interaction become more and more scarce.

In the regime with $d < 100$ nm, the deformation behavior might even change more drastically. Balancing the critical shear stresses for the nucleation of a full dislocation and a partial dislocation changes for the benefit of the partial if grain size is below a critical value [Chen et al., 2003]. This critical grain size is in the range of 10-15 nm for Al and 11-22 nm for Ni [Shan et al., 2004]. Furthermore, for polycrystalline microstructures, the GB volume fraction v_{GB} increases strongly, when grain size is in the lower nanometer regime; e.g. for 30 nm sized grains $v_{GB} = \delta A_{GB}/V_{GB} = 0.1$ is calculated, where A_{GB}/V_{GB} is the GB area per unit volume of a crystal, and $\delta = 1$ nm the GB thickness [Underwood, 1970]. Actually, for 10 nm sized grains even $v_{GB} = 0.3$ is computed. As a consequence, GB-mediated deformation could prevail over dislocation-mediated mechanisms. Possible processes are grain boundary sliding (GBS), grain rotation, or GB migration, carrying plastic deformation in an intragranular manner.

It is concluded that traversing several orders of magnitude of d, leads to changes in the dominant deformation mechanism: Coming from bulk-like behavior at the macroscale, individual dislocations, partial dislocations, and GB-mediated plasticity may dominate deformation with d approaching the nanoscale. Generally, geometrical constraints can be classified into dimensional (external dimension) and microstructural (internal dimension) nature [Arzt, 1998], which will be discussed in the following sections.

Furthermore, it is referred to Ref. [Kraft et al., 2010] for a comprehensive review on geometrical constraints.

2.1.1. Dimensional Constraint

Thin films are a prominent example for a structure with a dimensional constraint, since by definition their thickness is small compared to the other two dimensions. However, if the microstructural length scale (namely the grain size) is in the range of film thickness or even below, grain size rather than the film thickness dominates the mechanical response of the material [Venkatraman and Bravman, 1992; Thompson, 1993; Keller et al., 1998; Gruber et al., 2008a].

In order to investigate the pure effect of dimensional constraints, typically single crystalline pillars or wires are the chosen test geometries, to which a quasi-uniaxial load is applied. The characteristic length scale is in both cases the external geometry, namely the diameter. After very early experiments by Brenner on different FCC whiskers [Brenner, 1956, 1959], a renaissance arose in 2004 when Uchic et al. [Uchic et al., 2004] impressively showed the influence of the diameter of microcompression pillars on their strength. Microcompression pillar experiments are still a popular technique to investigate small scale mechanical behavior of single crystalline metals. The typical trends are that strength strongly increases with decreasing diameter and the stress-strain curve becomes serrated. The stochastic behavior is a consequence of individual slip events of dislocations and can be correlated to the formation of slip traces on the pillar surface.

It is obvious that the stochastic behavior strongly relies on the initial defect structure. However, most pillars are fabricated by FIB preparation, which becomes a serious issue for the smallest samples (pillar diameter $< 300\,\text{nm}$), since the volume affected by FIB-induced damage (e.g. Ga implantation), increasingly impairs the probed volume. As a consequence, it is impossible to accurately investigate defect-free, FIB-prepared pillars with smallest diameters. Based on a novel preparation route, defect-free one-dimensional structures, produced by chemically etching the matrix of directionally solidified NiAl-Mo eutectic [Bei et al., 2007], are obtained and thereby, the effect of defect density on mechanical behavior was proved without the undesired side effect of FIB-induced damage [Bei et al., 2008]: In their pristine state, these pillars with a diameter of ≈ 500 - $550\,$nm approach the theoretical yield limit under compressive loading. However, if defects are induced by predeformation the yield strength of similar test geometries decreases drastically, up to one order of magnitude.

For samples with small diameters, the tested volume, or respectively the gauge length, also plays an important role. This was investigated by experiments, comparing the compression behavior of pillars with low aspect ratios (small test volume) and the tensile behavior of fibers with high aspect ratios (large test volume) of the same Mo-based single crystalline material [Johanns et al., 2012]. As a result, the smallest pillars reach the theoretical strength of $\approx 10\,$GPa with negligible scatter, while for stretched fibers of the same diameter strength scatters over one order of magnitude from $1\,$GPa to $10\,$GPa. Similar to brittle ceramics, the different behavior can be explained by weakest-link concepts. The larger volume probed during a tensile test more likely contains an accumulation of defects, which, in this case, would lead to localized plasticity and thus, strongly reduced strength.

The size-dependent deformation behavior of single crystalline pillars can be summarized as follows: Bulk-like behavior in form of smooth stress-strain behavior and relatively low strength values is observed for the largest samples with several tens of micrometer in diameter. At the other end of the scale, the theoretical yield limit is reached for the smallest samples with $d \ll 1\,\mu$m, which are defect-free and behave predominantly elastic. In the interjacent transitional regime, serrated stress-strain behavior is observed resulting from slip events of individual dislocations. The stochastic defect distribution can cause a very large scatter in strength.

To gain more insight in the stochastic behavior observed in single crystalline structures, 3D Discrete Dislocation Dynamics (DDD) simulations [Weygand et al., 2002; Weygand and Gumbsch, 2005] are a suitable method, where the influence of dislocation densities and arrangements on the mechanical response can be deliberately examined [Senger et al., 2008; Motz et al., 2009; Senger et al., 2011].

An alternative to micropillar compression testing, is tensile testing of single crystalline fibers, as it was already done in the 1950's [Brenner, 1956, 1959]. With the advent of new fabrication routes [Richter et al., 2009] as well as novel and precise testing techniques [Gianola et al., 2011], defect-free wires with d as small as 50 nm can be tested accurately. Based on this combination, new deformation modes in Au nanowires can be unraveled [Sedlmayr et al., 2012]. A further alternative is tensile testing of singlecrystalline thin films supported by compliant substrate [Gruber et al., 2008c]. Generally, with the reduction of the characteristic length below ≈ 100 nm, the deformation behavior deviates from ordinary dislocation plasticity of full dislocations, and concomitantly partial dislocations as well as deformation twinning gain in importance.

2.1.2. Microstructural Constraint

Microstructural constraints in classical metallurgy, leading to enhanced strength, are particles, solute atoms, dislocations, or grain boundaries. Size effects are observed for all of them, if these barriers interact with dislocations. The relevant parameters for the occurrence of size-dependent behavior are particle, solute, and dislocation spacing, as well as grain size competing with the diameter of a dislocation loop [Arzt, 1998].

Pursuing grain refinement down to the nanometer regime, the trisection of the characteristic length, mentioned in section 2.1, is schematically shown in Fig. 2.1, which was adapted from Fig. 1 of Ref. [Kumar et al., 2003]. With ever-continuing decrease in grain size, the activity of full dislocations is more and more restricted. The classical Hall-Petch behavior must inevitably break down, if the grain size becomes smaller than the diameter of a dislocation loop. The oblate slope in the regime between 100 nm and 10 nm indicates that mechanisms different from ordinary dislocation plasticity may be active. The 'inverse' behavior for smallest grain sizes is plausible, since e.g. diffusive processes show a strong dependence on grain size ($\sigma \propto d^n$, where $n = 2\ldots3$) [Herring, 1950; Coble, 1963], acting in the opposite way than the Hall-Petch relation. In particular the stress required for activation of GB diffusion yields a $\sigma \propto d^3$ dependence [Coble, 1963]. In fact, decreasing hardness with decreasing grain size was rationalized by GB diffusion, active even at room temperature owing to the very small grain sizes ($D < 20$ nm) [Chokshi et al., 1989].

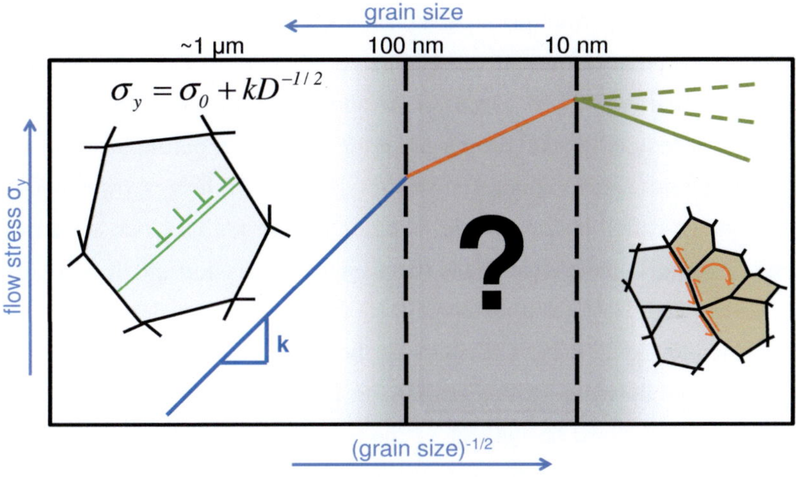

Figure 2.1.: Grain size dependence of yield strength, adapted from [Kumar et al., 2003]. No unified picture for deformation mechanisms in the regime of $10\,\text{nm} < D < 100\,\text{nm}$ exits. Moreover, for $D < 10\,\text{nm}$, different strength-grain size trends have been proposed.

2.2. Deformation Mechanisms in Nanocrystalline Metals

In this section, the current understanding of deformation mechanisms of NC metals is reviewed. For comprehensive reviews in literature it is referred to Refs. [Kumar et al., 2003; Meyers et al., 2006; Van Swygenhoven and Weertman, 2006]. The following review is subdivided into different mechanisms which were identified as being relevant in NC metals.

2.2.1. Dislocation Plasticity

Aggravating the confinement for dislocation multiplication, the activation stress for Frank-Read sources increases significantly [von Blanckenhagen et al., 2003]. Balancing the critical shear stresses for nucleation of a full dislocation and a partial dislocation changes for the benefit of the partial if grain size is below a critical value [Chen et al., 2003]. This critical grain size is in the range of 10-15 nm for Al and 11-22 nm for Ni [Shan et al., 2004]. Consequently, with decreasing grain size, rather partial dislocations are emitted from stress concentrations at GBs (e.g. GB ledges, triple lines). After the so-called leading partial is emitted from the GB, two scenarios can be imagined, dependent on the stress on the dislocation and the stacking fault energy (SFE, γ_{sf}) of the material: (i) The leading partial, which is still connected to the emitting GB traverses the grain and gets absorbed at the opposite GB, before a trailing partial can follow, leaving a stacking fault behind. (ii) If, based on the size of the grain and the SFE, the energy of the emerging stacking fault becomes too large, a trailing partial is emitted, following the leading partial. By combination of the two partials, a full dislocation is formed and the stacking fault dissolves [Yamakov et al., 2001; Derlet et al., 2003b]. If the leading and trailing partial are not split too much (high γ_{sf}), cross-slip can be an effective mechanism for a dislocation to overcome obstacles and to dissolve in the GB network [Bitzek et al., 2008].

MD simulations pointed to the relevance of the ratio of stable to unstable stacking fault energy for the dislocation behavior [Van Swygenhoven et al., 2004]. With a ratio close to 1, full dislocations can be expected, as splitting of the partials is constricted. On the other hand, for low ratios, extended partials should be dominant. Therefore, this ratio can - in contrast to the absolute value of the stable (or intrinsic) stacking fault energy - explain differences between different NC materials. However, conclusions from MD simulations on real material behavior must be treated with care, owing to the extremely high strain rates inherent to MD simulations. Indeed, also the authors ask the question how trailing partials would behave in a real experiment, where the time scale is not limited to nanoseconds.

From the experimental side, *in situ* XRD tensile testing of NC

Ni ($D \approx 26\,$nm) revealed reversible peak broadening [Budrovic et al., 2004], in contrast to ultrafine-grained (UFG) counterparts, prepared by high-pressure torsion (HPT), for which significant irreversible peak broadening was measured [Budrovic et al., 2005]. The reversible broadening was interpreted as evidence that no dislocation debris build up during deformation and that no substantial work hardening does occur. Instead, GBs might act as source and sink for lattice dislocations. The GB emission and absorption of dislocations would not leave additional dislocation debris, but only a modified GB network. On the other hand, the peak broadening observed for the HPT samples consists of reversible shares coming from elastic inhomogeneous strains and irreversible shares, arising from additional permanent dislocation networks. Hence, it is argued that if the grain size is small enough, conventional dislocation glide and multiplication are replaced by dislocations which are emitted from and absorbed in GBs.

Contrary to this interpretation, it was shown by sophisticated analysis of MD simulations [Bachurin and Gumbsch, 2010; Markmann et al., 2010], that dislocation activity is deferred to strains $> 3\%$, although additional irreversible microstrain emerges from the beginning of deformation. The initial increase in microstrain is therefore rather attributed to GB accommodation processes resulting from the heterogeneous response of the NC aggregate, and not to dislocations. Hence, the authors suggest to reconsider conclusions of previous *in situ* diffraction studies.

2.2.2. Deformation Twinning

Several recent studies point out the relevance of deformation twinning as one governing deformation mechanism in NC metals, see also Ref. [Zhu et al., 2012] for an overview. Partial dislocations are suspected to be responsible for the formation of deformation twins. In the former section, an argument, based on the nucleation stress of dislocations [Chen et al., 2003], was introduced, which benefits partial dislocation emission from GBs over full dislocation emission for grain sizes below $D \approx 10 - 20\,$nm. In this grain size range, deformation twinning was indeed identified by TEM [Chen et al., 2003] for Al, despite the high stacking fault energy of Al for which reason twinning was never identified in coarse-grained (CG) Al. Therefore it can be reasoned that a reduction of grain size, benefiting the nucleation of partial dislocations, can facilitate twinning. However, similar to CG metals, the occurrence of deformation twinning in NC metals strongly depends on the strain rate. It was shown in Refs. [Roesner et al., 2004; Markmann et al., 2003] that a small grain size is not the only prerequisite for twinning in NC metals, but high strain rates are required as well. Related to the high strain rates is the fact that the dominance of deformation twinning in NC metals was mostly identified by MD simulations (e.g. [Yamakov et al., 2002]) with inherently high strain rates, rather than by experimental observations.

In contrast to the above discussion of deformation twinning, in which the material was initially fairly free of twins, the mechanical behavior can also be modified by introducing a significant amount of growth twins to the microstructure during the fabrication process. Similar to grain boundaries, twin boundaries can impede dislocation motion. But especially coherent twin boundaries also bear a great potential for dislocation-twin interactions [Jin et al., 2006, 2008], which then can lead to very high strengths [Lu et al., 2009] and high strain hardening [Idrissi et al., 2011].

2.2.3. Grain Boundary-Mediated Plasticity

Many microscopic processes may contribute to GB-mediated plasticity. Maybe the most frequently discussed representative is grain boundary sliding (GBS). However, GBS requires accommodation processes for maintaining compatibility, e.g. diffusional processes or grain rotation. Only the combination of these processes adds up to a complete deformation mechanism. In contrast to dislocation-based plasticity, where each grain undergoes roughly the same shape change than the macroscopic sample (quasi-uniform flow), during GBS the grain shape remains the same (non-uniform flow) [Ashby and Verrall, 1973]. GBS is often associated with superplastic deformation, which classically involves high homologous temperatures. One approach to achieve large plastic deformations at RT is reducing the grain size to the NC range, which was proven to be successful [McFadden et al., 1999]. In particular for very small grains it was proposed that GB migration could also serve as accommodation process, which then could lead to (intergranular) mesoscopic glide planes involving several grains in length [Hahn et al., 1997]. Indeed, such glide planes were observed experimentally in Pd with $D \approx 14\,\text{nm}$ after severe plastic deformation [Ivanisenko et al., 2009]. Also MD simulations revealed the significance of GBS in samples with a NC structure. It was shown that atomic shuffling processes are involved in the GB region [Van Swygenhoven and Derlet, 2001]. Furthermore, the inverse Hall-Petch effect, which was related to the interfacial sliding model [Hahn et al., 1997] was also confirmed by MD simulations [Schiotz and Jacobsen, 2003]. This means a maximum strength was found in the range of $D = 10\text{-}20\,\text{nm}$, and reduced strength for smaller grain sizes.

The processes discussed so far in this section do not enforce any behavior related to the crystallographic grain orientation, since not the crystallographic orientation matters, but the local configuration of the interfaces. Therefore, it is often argued that these mechanisms do not yield a crystallographic texture after deformation, but rather randomize grain orientations [Ma, 2004]. On the other hand, if grain rotation is not accompanied by GBS or diffusional processes, but dislocations accomplish a grain rotation, a preferred crystallographic orientation emerges and concomitantly a change in grain shape. Hence, applying texture analysis (e.g. [Markmann et al., 2003; Fan et al., 2006a]) can often help to differentiate between dislocation- and GB-mediated plasticity.

Excessive grain rotation can lead to multigrain agglomerates, which was observed during *in situ* TEM experiments on NC Ni films [Shan et al., 2004; Wang et al., 2008]. Thereafter, analysis of the substructure of the grain agglomerates have shown that the agglomerates consist of several subgrains with a special edge-on orientation.

Although, it is challenging to measure the relative contribution of GB-mediated deformation to overall deformation, the recently observed shear softening of GBs in NC Pd [Grewer et al., 2011] supports the idea that GB-mediated shear and slip [Weissmueller et al., 2011] may carry a dominant share of macroscopic deformation, especially at stress levels too low to mobilize (partial) dislocations.

2.2.4. Grain Boundary Migration

Recently, the significance of GB migration as a distinct deformation mechanisms was proclaimed [Gianola et al., 2006; Rupert et al., 2009]. Shear stress driven GB migration leads to inhomogeneous grain growth, where some grains preferentially grow at the expense of others. A maintained initial grain shape and an increased standard deviation of the grain size distribution are characteristic for the tested microstructures. When the size of individual grains is significantly increased, ordinary dislocation-based deformation can take place, leading to an extended plastic behavior [Gianola et al., 2006]. Certainly, the relevance of GB mobility is dependent on the constitution of the GBs: If the GBs are successfully pinned, e.g. by impurities, this mechanism can be easily switched off [Gianola et al., 2008]. Based on an analytical model, it is argued that the relative grain translation is coupled to a normal GB motion

[Cahn et al., 2006]. The geometric coupling factor β relates the absolute normal motion to the tangential motion. In that way, GB migration can produce plastic strains.

It is noted that all above mentioned experiments were conducted on freestanding thin films. One can question, how strong the effect of anomalous growth of individual grains is related to the thin film geometry involving many free surfaces and a high surface-to-volume ratio, and how relevant the effect is for bulk samples.

2.2.5. Alloying Effects

The stabilizing effect of foreign atoms on as-fabricated NC structures are quite well understood [Koch et al., 2008]. Solute atoms have been proven to be effective in reducing the initial grain size by solute-drag [Wang et al., 2007]. However, the effect of alloying on deformation mechanisms of NC structures has only received little attention. Though, it was shown in recent studies [Scattergood et al., 2008; Schaefer et al., 2011] that alloying can have a major effect on the deformation behavior. In order to focus on the relevant topics for this thesis, the discussion is restricted to alloys forming a solid solution. Generally, a solid solution strengthening effect strongly depends on the misfit of the atomic radii of matrix atom and solute, as well as, on the differences in shear moduli, affecting both the mobility of dislocations [Labusch, 1970]. On the other hand, alloying can significantly change the material properties, e.g. stacking fault energy, which is an important factor for the deformation behavior in the NC regime (cf. section 2.2.1). It was shown in Ref. [Schaefer et al., 2011], that the intrinsic stacking fault energy can be continuously reduced from $\gamma_{sf} \approx 180\,\mathrm{mJ/m^2}$ for pure Pd to less than a forth for very high solute concentrations of Au, which in turn significantly increases the stacking fault density during deformation. It is argued that the reduction of γ_{sf} leads to a wider splitting of leading and trailing partial dislocation, and thereby extended partials could dominate full dislocations.

2.2.6. Summary of Deformation Mechanisms in the NC Regime

It is clear that due to the geometrical confinement of small grain sizes, the conventional deformation mechanisms of dislocation glide, multiplication, pile-up, and interaction are constrained in the NC regime. In the microcrystalline and ultra fine grained regime, the impediment leads to a strengthening effect with basically maintaining the conventional mechanisms. At the other end of the scale, for the smallest grain sizes ($D \approx$ 5 nm), GB-mediated deformation, e.g. GBS accommodated by rotation or diffusive processes, seems to dominate the deformation behavior. However, in the regime in between several different deformation mechanisms may be active, either successively as a function of the applied strain and/or simultaneously. Besides the grain size, there are lots of effects which can influence the occurrence and domination of individual mechanisms (e.g. strain rate, impurities, texture, grain size distribution, etc.). This complicates comparison between individual studies, and averts a more unified picture of deformation mechanisms in NC metals.

The state-of-the-art discussed above is illustrated schematically in Fig. 2.1, which is based on Fig. 1 of Ref. [Kumar et al., 2003]. For $D > 100$ nm, classical HP strengthening is observed. On the other hand, for $D < 10$ nm often an inverse HP behavior is observed, where GB-mediated deformation likely becomes dominant. In the intermediate regime, 10 nm $< D < 100$ nm, the slope is oblate, allowing for speculations on different mechanisms and different scaling behavior.

The boundaries between the regimes are not very well defined and an abrupt crossover between the regimes, as displayed in the original figure, seems rather unlikely. Therefore, the gradually increasing impediment of ordinary dislocation plasticity and Hall-Petch strengthening is illustrated with a blurred crossover around 100 nm. Also the crossover at very small grain sizes is blurred, since different critical grain sizes ranging from $D \approx 10$ - 30 nm have been found ([Schiotz and Jacobsen, 2003; Chen et al., 2003; Trelewicz and Schuh, 2007]). Below the critical grain size, alternative trends in the strength-over-grain size diagram are indicated, based on recent results [Trelewicz and Schuh, 2007; Meyers et al., 2006]. While often an inverse Hall-Petch behavior is discussed, also plateaus with strength values close to the theoretical strength were found [Trelewicz and Schuh, 2007]. Overall, so far no consensus is reached on the mechanical behavior of metals at the lower end of the NC regime. In particular, a gradual crossover between deformation mechanisms is likely, which would imply the coexistence of different mechanisms and broaden a single critical grain size to a rather vague grain size range.

2.3. Open Questions and Aims of this Study

In the reviewed literature (e.g. [Chen et al., 2003; Budrovic et al., 2004; Shan et al., 2004; Rupert et al., 2009]), mostly one individual deformation mechanism is discussed, however it is speculated that different mechanisms operate in the NC regime, especially in the crossover regime indicated in Fig. 2.1, either simultaneously and/or successively. So far, it was - to the author's knowledge - not possible to experimentally identify and separate the relative contributions of the different mechanisms and to allocate them to specific stress and strain regimes.

An interesting attempt to separate dislocation- and GB-mediated deformation has been made by MD simulations during uniaxial compression testing [Vo et al., 2008]. The relative shares were evaluated dependent on the applied strain, grain size, and strain rate. In fact, the strain rate dependence showed a serious impact on the relative shares of both types of mechanisms to overall deformation. Thus, due to the strong strain rate dependence and the overall exorbitant high strain rates inherent to MD simulations, conclusions on real material behavior are limited and still lacking.

The relative shares are also significantly influenced by the applied strain. Other MD simulations have shown that dislocations are first observed for strains larger 3% [Bachurin and Gumbsch, 2010] or even larger 4% [Bitzek et al., 2008]. Furthermore, it is discussed that deformation mechanisms do not leave a footprint after deformation (e.g. [Budrovic et al., 2004]). Consequently, in order to analyze and identify deformation mechanisms, which might be reversible or change with the degree of deformation, *in situ* investigation methods are required. In recent years, mainly two *in situ* methods were established for the investigation of deformation mechanisms in NC materials: *In situ* TEM and *in situ* XRD. *In situ* TEM allows for a straightforward interpretation of observed microstructural changes. However, it always suffers from limitations on (i) preparation-induced damage to the sample and (ii) surface effects of the only several nm thin foils, making conclusions of the bulk behavior sometimes questionable. In contrast, *in situ* XRD, using 3rd generation synchrotron sources with high brilliance in combination with fast detectors, samples over a large volume and provides insight on the average material behavior [Van Swygenhoven et al., 2006]. *In situ* XRD has the advantages of excellent time resolution and statistical relevance. By using large area detectors, texture evolution can be monitored during deformation, which can help to differentiate between dislocation- and GB-mediated deformation [Ma et al., 2004].

An extended microplastic regime - which is defined as a transitional regime between purely elastic and macroplastic deformation - was reported for NC materials [Saada, 2005; Brandstetter et al., 2006]. This microplastic regime defers fully plastic behavior to higher strains, as also discussed in the former paragraph. However, tensile experiments on NC bulk metals, which have been mostly conducted, are limited to strains of a few percent (6.5% [Budrovic et al., 2004], 7.2% [Fan et al., 2006a]). Maximum strains for other tensile studies are even lower ($< 3\%$, e.g. [Fan et al., 2009]). As a consequence, the macroplastic regime, especially $\varepsilon > 10\%$, is pretty much unexplored.

Since the deformation behavior of NC metals is still under heavy debate, especially in the grain size range between 10 nm and 30 nm where a crossover in mechanisms is predicted and expected [Schiotz and Jacobsen, 2003; Trelewicz and Schuh, 2007; Chen et al., 2003], this thesis aims to shed light on the complexity of deformation mechanisms in NC metals, particularly in the crossover regime. The identification of active deformation mechanisms, and the separation and classification to specific stress and strain regimes are in the main focus. Thereby, the sequence of different mechanisms, and their possible coexistence, especially in the unexplored macroplastic regime, should be untangled.

These objectives are pursued with an *in situ* XRD technique during mechanical testing. The principle is based on a setup for measuring the evolution of biaxial lattice strains and stresses in thin films under tensile load [Bohm et al., 2004] and is upgraded in the following way:

- Due to the high accuracy of the data, not only the most stable peak parameter, namely the peak position, is tracked. Instead, a complete and sophisticated peak shape analysis is performed incorporating asymmetric peak profiles, additionally investigating the evolution of peak broadening, asymmetry and intensity during mechanical testing.

- Two approaches are pursued in this thesis to explore the macroplastic regime: (i) Compression experiments and shear-compression experiments are conducted on NC bulk metals and alloys, allowing for large plastic strains, despite the inherent limited ductility. (ii) Owing to the use of compliant substrates, deformation of thin films can be delocalized [Xiang et al., 2005], and large plastic deformation ($\varepsilon \approx 30\%$) without film cracking can be accomplished, even for metals and alloys with a NC microstructure owing to improved preparation techniques [Lohmiller et al., 2010].

- Complete Debye-Scherrer rings are recorded with area detectors and analyzed by radial scans every $2°$. This allows for monitoring the evolution of deformation textures during loading. Moreover, by analysis of peak broadening and its in-plane dependent behavior, the grain size and shape can be analyzed, respectively.

The investigated NC metals are Nickel (Ni), Palladium (Pd), and Palladium-Gold (Pd-Au) alloys. Nickel was often used as a reference for FCC NC metals. Numerous literature exists, which can serve for comparison. The investigation of Pd and PdAu alloys is embedded in a large research unit (FOR714) of the German Research Foundation (DFG), also providing a large database for comparison. In particular, grain sizes ranging from 10 nm to 30 nm are investigated, in order to elucidate the interplay of different deformation mechanisms. For the classification into different deformation regimes, several relevant parameters, including lattice strain, texture, grain size, and microstrain, will be considered.

All tested samples were fabricated in close collaboration with cooperation partners, ensuring complete control of all preparation steps, which can have a significant influence on microstructure and hence on mechanical behavior. Between fabrication and mechanical testing, detailed 3D microstructural characterizations were conducted, allowing to establish microstructure-property relationships.

The *in situ* XRD investigations are supported by various other characterization techniques, including *in situ* testing under the optical microscope and under the scanning electron microscope (SEM), as well as, *ex situ* characterization by transmission electron microscopy (TEM) in combination with Automated Crystal Orientation and phase Mapping (ACOM), as well as, conventional XRD and SEM.

3. Experimental

3.1. Introduction

The main experimental technique of this thesis is synchrotron-based X-ray diffraction (XRD) during mechanical testing. Dependent on sample type and geometry, different loading conditions were applied to the NC metallic samples: (i) Tensile testing on thin films with a typical thickness of $\approx 1\,\mu m$ adherent to compliant substrate, and (ii) compression testing and shear-compression testing on bulk samples with typical sample dimensions in the mm range. The high flux of 3rd generation synchrotron radiation facilities provides excellent time resolution in comparison with laboratory setups, i.e. all experiments are conducted continuously avoiding stepwise straining, where samples would suffer from relaxation effects during XRD measurements after each strain step. Bragg's Law constitutes the basic principle of XRD:

$$\lambda = 2d_{hkl}\sin(\theta_{hkl}) \tag{3.1}$$

where λ is the X-ray wavelength, d_{hkl} the lattice spacing of specific (hkl) planes and θ_{hkl} the corresponding Bragg's angle, which is half of the (hkl) peak position. Powder diffraction (monochromatic X-rays, $\lambda = const.$) is applied, since all tested samples consist of a polycrystalline microstructure. Due to the constant wavelength, changes in lattice spacing, as a result of the applied load, are monitored by shifting Bragg reflections. In theory, each reflection of a perfect crystal only consists of a single Dirac delta function. However, in reality the reflections are always broadened due to instrumental effects and the microstructure of the sample. The instrumental effects can be taken into account by deconvolution of the broadening measured for the relevant sample and a reference measurement of a standard powder sample, e.g. LaB_6. The actual net broadening caused by the sample under investigation can be separated into size- and strain-dependent shares with well-established methods, such as Warren-Averbach (WA) [Warren and Averbach, 1950], Williamson-Hall (WH) [Williamson and Hall, 1953], or Single Line Method (SLM) [de Keijser et al., 1982]. This allows for calculating the coherent scattering domain size (mostly referenced as grain size) and microstrain. Microstrain results from inhomogeneous strain fields and broadens the reflections, while homogeneous strain, correlating with changes in lattice spacing, does not broaden the reflections, but shifts the reflection in a self-similar manner. Reasons for inhomogeneous strain can be grain-to-grain interactions or - on a more local scale - strain fields around dislocations, stacking faults, or other lattice defects.

3.2. Setups for Synchrotron-Based In Situ XRD Mechanical Testing

The experiments presented and discussed in this thesis were performed at different beamlines in order to meet the demands of the different sample types and geometries. In principle, all setups are based on a setup originally developed at the MPI-MF Surface Diffraction beamline at ANKA in order to measure the evolution of biaxial lattice strains and stresses in thin films under load [Bohm, 2004; Bohm et al., 2004]. The individual setups and their characteristics will be explained in detail, but first, the common characteristics are denoted in the following listing and also can be seen in the schematics of Fig. 3.1:

- The sample surface is oriented perpendicular to the incoming monochromatic X-ray beam. This means, that the scattering vectors of different diffracting (hkl) planes are oriented differently with respect to the loading direction. A further consequence is, that the in-plane information of a sample is probed, in contrast to classical Bragg-Brentano measurements where in fact the out-of plane information is probed.

- X-ray detectors (preferentially an area detector or two line detectors) are placed in transmission geometry.

- The combination of cross section of the X-ray beam and thickness of the illuminated sample provides a large sampled and uniformly deforming volume yielding excellent statistics.

- The total strain is measured on the sample surface with an optical camera and calculated by Digital Image Correlation (DIC) or Feature Tracking [Eberl, 2010].

- The mechanical testing device (Kammrath and Weiss, Germany) allows applying tensile and compressive loads up to $10\,kN$ with velocities ranging from $0.15\,\mu m/s$ to $20\,\mu m/s$. Since the testing device is mounted on the diffractometer, easy sample positioning, with respect to the incoming beam, is possible.

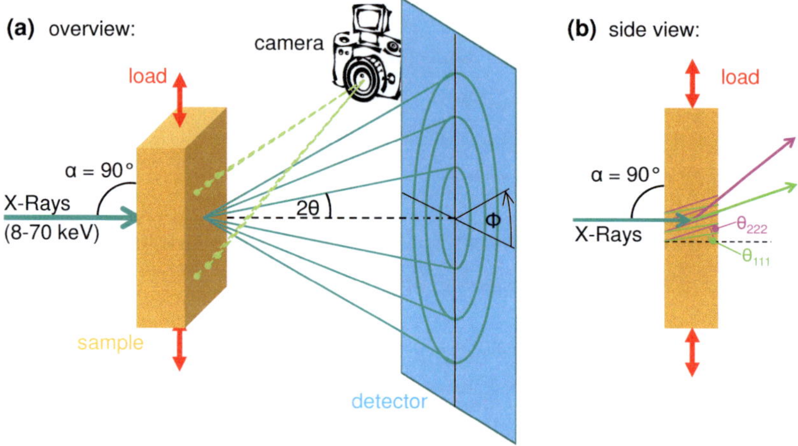

Figure 3.1.: (a) Overview of the principle synchrotron-based *in situ* XRD mechanical testing setup with transmission geometry: Energy of the X-ray beam ranges between $E_{Xray} = 8\text{-}70\,$keV. The radial position of a reflection is given by the angle 2θ and the in-plane position by the azimuthal angle ϕ. Tensile and compressive loads can be applied to different sample types. A camera system allows to track the applied strain. (b) Side view: An important characteristic of the setup with transmission geometry is, that the diffracting lattice planes are oriented differently with respect to the loading direction. The higher the indices of the diffracting planes, the higher is the angle between plane normal and loading direction, which corresponds to the Bragg's angle θ. As a consequence, the diffracting planes experience reduced load with increasing scattering angle 2θ.

3.2.1. High Energy In Situ XRD Setup at the ESRF ID15A for NC Bulk Samples

Experiments on NC bulk samples were carried out at the High Energy Microdiffraction (HEMD) endstation of beamline ID15A of the ESRF, France. With the following setup, also described in [Lohmiller et al., 2012b], pure compression tests and miniaturized shear-compression tests were conducted on different NC bulk materials at room temperature. An overview of the setup is presented in Fig. 3.2.

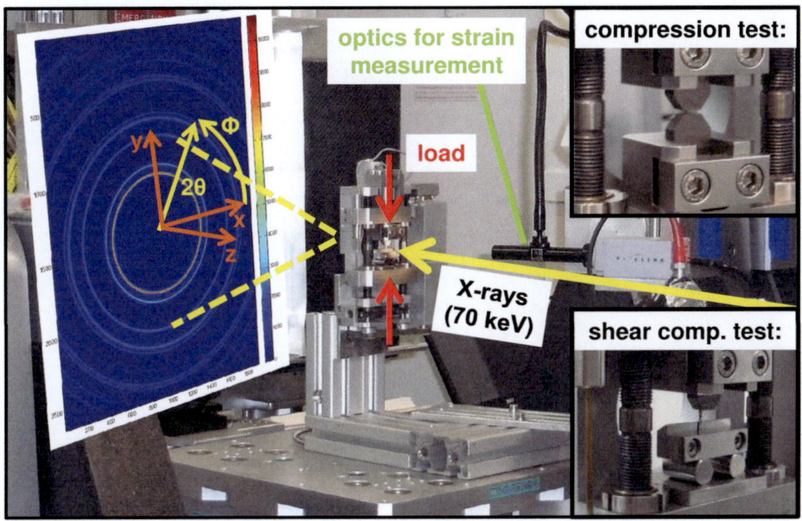

Figure 3.2.: High energy *in situ* XRD setup at the beamline ID15A of the ESRF for testing NC bulk samples with thicknesses in the mm range. The arrangement for compression tests is shown in the upper inset and for shear-compression tests in the lower inset.

Bulk samples with a thickness of $\approx 1\,$mm can be penetrated with the high energy of $E_{Xray} = 70\,$keV. The high energy corresponds to a wavelength $\lambda = 0.1776\,$Å. As a result, the diffracting planes are almost parallel to the incoming X-ray beam, e.g. $\theta_{111,Ni} = 2.5°$. The beam size was $20\,\mu$m in horizontal and $8\,\mu$m in vertical direction. As detector either a Pixium 4700 flat-panel detector (Thales Electron Devices, Moiron, France) [Daniels and Drakopoulos, 2009] or a $165\,$mm MAR-CCD (MAR, Inc., Evanston, IL, USA) was used. The main characteristics of both area detectors are summarized in Table 3.1.

With this setup, complete Debye-Scherrer rings from the (111) up to the (222) reflection were recorded. The fast detectors and the high X-ray flux allowed for continuous tests with strain rates in the order of $10^{-4}\,$s^{-1}. The

Table 3.1.: Comparison of two area detectors at the ESRF ID15A.

Detector	Pixium 4700	MAR 165
Active area	$382\,mm \times 294\,mm$	$\varnothing 162\,mm$
Pixel array	2480×1910	2048×2048
Pixel size	$0.15400\,\mu m$	$0.07894\,\mu m$
Distance to sample	$\approx 81\,cm$	$\approx 46\,cm$

arrangement for pure compression tests with two hard metal punches is shown in the upper inset of Fig. 3.2, whereas for shear-compression tests it is shown in the lower inset.

For the compression experiments a typical sample geometry of $\approx 1 \times 1 \times 0.6\,mm^3$ was used. The total compressive strain ε_c was measured by a CMOS camera (PixeLINK PL-B782U, USA) with a $6\times$ in-line illumination telecentric lens (Edmund Optics NT59-743, USA) and calculated by DIC. A homogeneous grid of markers, tracking the relative displacement, was distributed over the sample in the recorded image sequence. In post-processing unreliable markers were deleted.

For the experiments of shear-compression specimens (SCS), the lower punch is replaced by a sample holder borne on two metal rolls allowing for lateral movement of the lower sample half as a result of shear deformation in the sample slit. The typical SCS test assembly is schematically shown in Fig. 3.3(a). The very small beam size of the beamline ID15A is required to probe the materials response by XRD in the narrowed slit of the sample (width and height $\approx 100\,\mu m$) where shear is prevalent. Furthermore, the high energetic X-rays are required yielding low 2θ diffraction angles and thereby allowing the diffracted beam to exit the slit without penetrating the bulk. For details of the nonstandard shear-compression tests it is referred to Ref. [Ames et al., 2010].

The SCSs were compressed with a constant velocity of $1\,\mu m/s$. The macroscopic deformation of the slit was assessed by DIC analysis of the

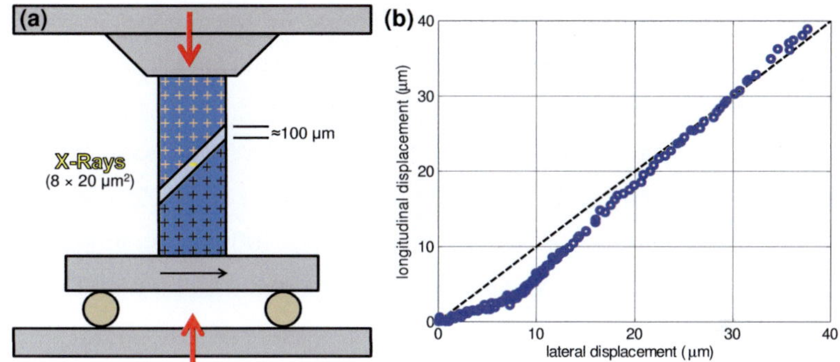

Figure 3.3.: (a) Sketch of the SCS in the load frame. The two metal rolls allow for lateral movement of the lower sample half, as a response of shear in the slit. The crosses indicate the marker grids of the DIC analysis. Note that the dimensions are not true to scale. (b) Evaluation of the relative displacements of the upper to the lower sample half in longitudinal and lateral direction, analyzed by DIC. Longitudinal equals almost lateral displacement, thereby indicating dominant shear deformation.

relative movement of the rigid upper to the lower bulk sample half in longitudinal and lateral direction (see the marker grids in Fig. 3.3(a)). The results are shown in Fig. 3.3(b). Besides some deviation at the beginning of loading, which can be attributed to settlement effects, the longitudinal displacement is similar to the transversal displacement, indicating macroscopic shear.

3.2.2. Ultra Fast In Situ XRD Setup at the SLS MS04 for NC Thin Films

Polyimide-supported thin films were tensile tested with the following setup at the MS04 powder beamline of the SLS, Switzerland. A photograph of the setup is shown in Fig. 3.4 and a detailed description is found in Refs. [Lohmiller et al., 2012a; Olliges et al., 2007]. A large solid-state Mython microstrip line detector [Schmitt et al., 2003], which covers an angular range of $-60° < 2\theta < 60°$, was placed in transmission geometry parallel to the tensile axis, and a second detector module in lateral direction. The energy of the X-ray beam was chosen to be $E_{Xray} = 17.5\,keV$, which translates to a wavelength of $\lambda = 0.71\,\text{Å}$. Reflections from (111) to (333) were recorded parallel to loading and from (111) to (222) in lateral direction. The beam size was approximately $500 \times 500\,\mu m^2$, probing a large area of the uniformly deforming substrate-supported metal films. The fast readout of the detectors allows to deform the specimens continuously with strain rates in the range from $10^{-6}\,s^{-1}$ to $10^{-4}\,s^{-1}$, and concomitantly to attain excellent signal-to-noise ratios. The true strain ε was measured with a SLR camera (Nikon D80, Japan) equipped with a 200 mm macro-objective (Nikon, Japan) and calculated by Feature Tracking. Both, the X-ray diffraction patterns and the total strain, were measured locally in the center of the sample.

3.2.3. In Situ XRD Setup with 2D-Detector at the ANKA MPI-MF for NC Thin Films

In order to investigate in-plane effects of the thin film samples adherent to polyimide substrate the following setup equipped with a 165 mm MAR-CCD detector was used at the MPI-MF beamline at ANKA, Karlsruhe. The detector is similar to the MAR-CCD used in the setup for bulk samples introduced in section 3.2.1. An overview of the setup is presented in Fig. 3.5.

To account for the pronounced <111> fiber texture of the primarily tested Au-based thin films, the energy was adjusted to $E_{Xray} = 7.97\,keV$ ($\lambda = 1.56\,\text{Å}$) [Bohm et al., 2004]. In this setup, the area detector recorded the complete (111) and (200) Debye-Scherrer rings. Beam size was approximately $500 \times 500\,\mu m^2$. The tensile tests were performed with strain rates in the order of $10^{-5}\,s^{-1}$. The true strain ε is measured by a CMOS

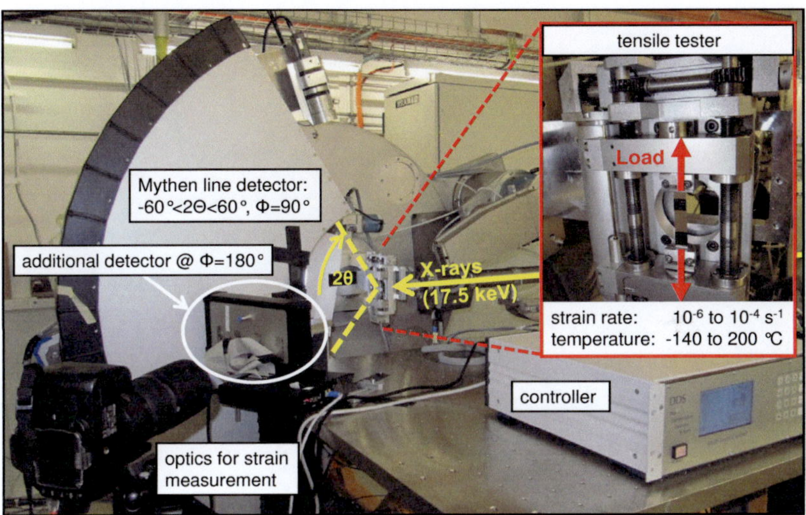

Figure 3.4.: *In situ* tensile testing setup at the MS04 beamline of the SLS. The ultra fast and large MYTHEN microstrip detector is placed in transmission geometry parallel to the loading direction, and covers an angular range $-60° < 2\theta < 60°$. In lateral direction additional detector modules are positioned. The inset shows the mechanical testing device, including a mounted sample on which the speckle pattern for Feature Tracking is visible.

camera (PixeLINK PL-B782U, USA) equipped with a $1\times$ telecentric lens (JENmetar $1\times$/12 LD, JENOPTIK, Jena) and calculated by Feature Tracking using a speckle pattern around the X-ray illuminated area (see inset of Fig. 3.5).

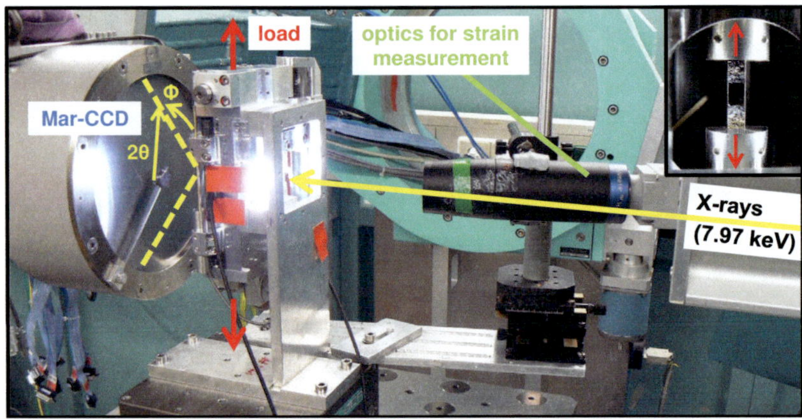

Figure 3.5.: *In situ* tensile testing setup at the MPI-MF beamline at ANKA, Karlsruhe equipped with an area detector in order to investigate in-plane effects of thin film samples. The inset shows a sample with a speckle pattern which is used to determine true strain.

3.3. Data Analysis of Diffraction Patterns

The evolution of XRD diffraction patterns is described in terms of peak parameters, such as position, width, shape, and intensity of any recorded (hkl) reflection along any in-plane direction ϕ, and how they evolve during mechanical testing. In the following section, it will be explained how these peak parameters are obtained starting from the recorded diffraction patterns.

3.3.1. Processing of Diffraction Patterns and Background Subtraction

The setups at ESRF (section 3.2.1) and ANKA (section 3.2.3) are equipped with area detectors. For the data analysis, radial scans are cut out from the complete Debye-Scherrer rings at a certain ϕ angle and with a radial width of $\Delta\phi = 0.4°$, as shown in Fig. 3.6(a). This is repeated for all ϕ angles with increments of $2°$. The resultant I vs. 2θ-scans (Fig. 3.6(b)) are similar to what is obtained from a line detector for a single ϕ direction, as used in the SLS setup (section 3.2.2).

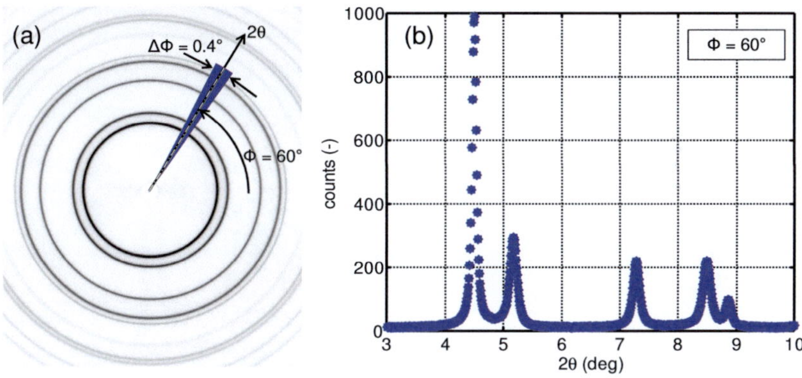

Figure 3.6.: Diffraction pattern of a NC Pd-70Au sample ($D \approx 10\,\text{nm}$) recorded at beamline ID15A of the ESRF. (a) Complete Debye-Scherrer rings from (111) to (222). Radial scans are cut out every $2°$ and averaged over $\Delta\phi = 0.4°$, resulting in an (b) I vs. 2θ-scan for each ϕ angle. Please note the overall low background level.

The as-recorded diffraction patterns contain, besides the structural information of the sample under investigation, diffuse scattering and instrumental effects. The diffuse scattering primarily arises from air scattering.

In the case of bulk samples, the hereafter introduced fit model includes a 1st order polynomial, taking the diffuse scatter into account which decreases with increasing 2θ. Since the fit model is only applied to single freestanding or two overlapping reflections, this approach is appropriate for the narrow 2θ range. The overall low background level can be seen in Fig. 3.6(b).

In the case of polyimide-supported thin films, a halo from the amorphous substrate is additionally superimposed to the overall signal. This is taken into account with the subsequent procedure, which is also presented in Fig. 3.7: (i) A blank measurement of the pure substrate is recorded, which also includes the air scattering (red); (ii) for a defined 2θ range, a spline is modulated to this reference measurement (black); (iii) the spline is subtracted from the recorded data (blue), resulting in the net structural information of the selected data range (green).

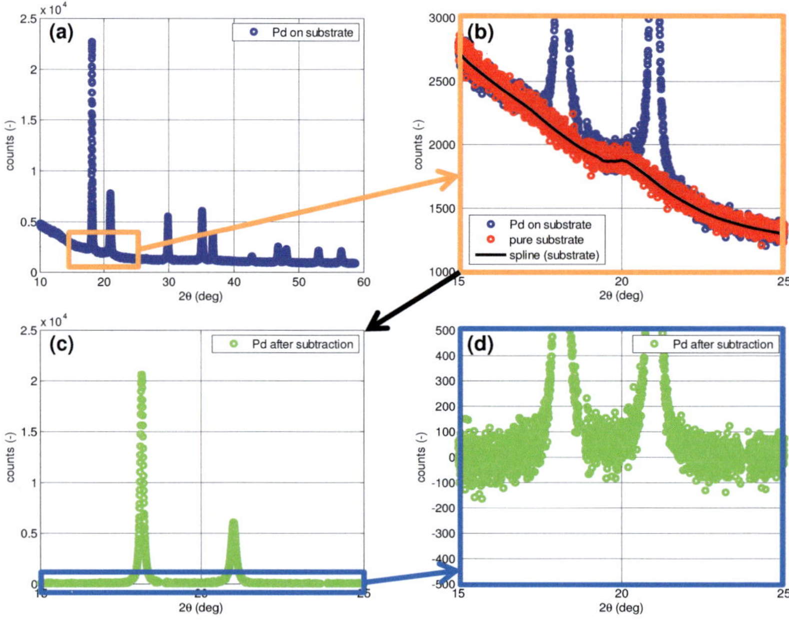

Figure 3.7.: From the as-recorded data to the net structural information, exemplarily shown for a polyimide-supported Pd thin film measured at the SLS: (a) The as-recorded pattern of the film comprises 10 peaks and includes all background. (b) A blank measurement of the pure polyimide substrate is recorded, which also includes the air scattering. In a defined 2θ range, a spline is modulated to the background signal. (c, d) The spline is subtracted from the recorded data, resulting in the net structural information. Please note the excellent signal-to-noise ratio.

Instrumental effects primarily affect the broadening of the reflections. Therefore, for further calculations on the basis of peak broadening, such as grain size - microstrain separation, the instrumental broadening is taken into account with a standard measurement of a reference powder (Y_2O_3 and CeO_2) and the subsequent deconvolution of the sample and the standard measurement (see section 3.4.2).

3.3.2. Peak Fitting

In order to describe the peak shape as accurate as possible, a very flexible peak fit model is required. Since the reflections are neither Gaussian nor Lorentzian shaped, a function with a mutable peak shape factor is required. Furthermore, the reflections may develop a pronounced asymmetry during loading, which also has to be considered in the model. Therefore, an asymmetric split Pearson VII function [Toraya, 1990] was used. The necessity for the use of such a flexible peak fit model is demonstrated in Fig. 3.8.

The peak fit function for a single reflection without background is given by

$$P(2\theta) = A_0 \left(H(2\theta_B - 2\theta) \left[1 + \left(\frac{1+A_{P7}}{A_{P7}} \right)^2 \frac{(2\theta - 2\theta_B)^2 (2^{1/\mu} - 1)}{\omega^2} \right]^{-\mu} \right.$$

$$\left. + H(2\theta - 2\theta_B) \left[1 + \left(\frac{1+A_{P7}^{-1}}{A_{P7}^{-1}} \right)^2 \frac{(2\theta - 2\theta_B)^2 (2^{1/\mu} - 1)}{\omega^2} \right]^{-\mu} \right) \qquad (3.2)$$

where A_0 is the maximum peak intensity or peak amplitude, H a Heaviside function, 2θ the position of the peak maximum, A_{P7} the peak asymmetry parameter, μ the peak shape parameter differentiating between Lorentzian ($\mu = 1$) and Gaussian ($\mu \to \infty$) peak shape and ω the full width at half maximum (FWHM). For convenience, the intrinsic asymmetry parameter of the Pearson VII function, A_{P7} , is converted into the alternative form

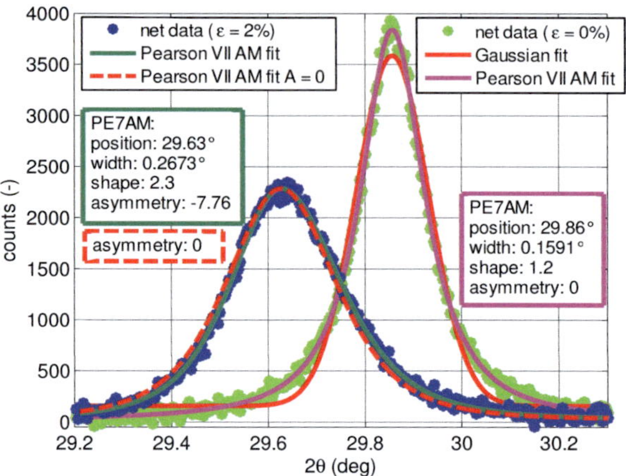

Figure 3.8.: Example of a Pd thin film tested at the SLS that shows the necessity for a flexible peak fit model. Unstrained data (green dots), which is inadequately represented by the Gaussian fit (red line), but reasonable by the PE7AM fit (magenta line). Under load a pronounced asymmetry emerged (blue dots), which is taken into account by the PE7AM fit (green line). Especially in the shoulders of the reflection, the discrepancy between asymmetric and symmetric (red dashed line) Pearson VII fit functions is remarkable.

$A = (A_{P7} - 1)/(A_{P7} + 1) \times 100\,(\%)$ in order to yield equal absolute values for positive and negative skewness of the peak. Furthermore, by integration, the integral peak intensity (INT), i.e. peak area and the integral peak breadth (IBR), i.e. peak area divided by peak amplitude, can be obtained.

3.4. Further Data Evaluation

Often it can be very useful to continue processing the obtained peak parameters, to gain more insight in the microstructure and how microstructure evolves during loading. In the following, calculation of the elastic lattice strain from the peak position and a method to extract grain size and microstrain from the peak broadening during loading will be motivated and explained.

3.4.1. Elastic Lattice Strain

The elastic lattice strain can be easily calculated from the peak position. The evolution of lattice strain over total strain helps to establish a classification into elastic and plastic deformation behavior. Furthermore, different alloy compositions in continuously miscible alloy systems have a significant influence on the initial peak position due to different lattice constants. In order to compare the peak shifts of different alloys, the conversion of peak position to lattice strain is helpful. The (hkl) lattice strain for a specific azimuthal direction ϕ is calculated according to

$$\varepsilon_{hkl,\phi} = \frac{d_{hkl,\phi} - d^0_{hkl,\phi}}{d^0_{hkl,\phi}} \tag{3.3}$$

where $d_{hkl,\phi}$ is the (hkl) lattice spacing in a specific azimuthal direction ϕ. The superscript '0' denotes the initial state. With the help of Equation (3.1) it follows

$$\varepsilon_{hkl,\phi} = \frac{\sin(\theta^0_{hkl,\phi})}{\sin(\theta_{hkl,\phi})} - 1 \tag{3.4}$$

where $\theta_{hkl,\phi}$ is the (hkl) Bragg's angle in a specific ϕ direction. Note that, given by the transmission geometry, the considered planes are inclined by θ with respect to the incoming X-ray beam. Likewise, the plane normals are inclined by θ with respect to the loading direction (cf. Fig. 3.1(b)). As a consequence, with increasing 2θ angle, the diffracting planes experience lower stresses. This means, the (222) planes yield generally lower lattice strains than (111) planes, although they are equally elastically compliant. To get a rough idea how the inclination angles vary between the different setups, based on the different X-ray energies used, the Pd (111) Bragg's angle $\theta_{111,Pd}$ is denoted for the three setups, measuring $\approx 2.2°$ for the ESRF setup ($E_{Xray} = 69.7\,\text{keV}$, section 3.2.1), $\approx 9°$ for the SLS setup ($E_{Xray} = 17.5\,\text{keV}$, section 3.2.2) and $\approx 20°$ for the ANKA setup ($E_{Xray} = 7.97\,\text{keV}$, section 3.2.3). Finally, it is pointed out that the lattice strain value is only a relative measurement with respect to the initial state, and d^0 may slightly vary from the strain-free lattice spacing due to residual stresses after sample fabrication.

3.4.2. Grain Size and Microstrain

Grain size D and microstrain $<\varepsilon>$ of the sample microstructure can be obtained by separating size- and strain-dependent broadening shares. Different X-ray line broadening analyses are possible. A compact overview of different methods is given in Ref. [Mittemeijer and Welzel, 2008]. For the *in situ* studies presented in this thesis, the relatively simple Single Line Method (SLM) [de Keijser et al., 1982] was chosen, in order to avoid complications due to the complex elastic grain interaction during loading within the NC microstructure. Furthermore, since the incoming beam penetrates the sample perpendicularly, all scattering vectors are oriented differently with respect to the loading direction (which is not the case for the Bragg-Brentano geometry which is usually used for this kind of analysis), and therefore the different planes experience different loads. This becomes evident by distinct differences of the calculated lattice strains for any (hkl) planes and their higher orders (e.g. (111) and (222)), even with the high energy setup, which should be similar if the deviation of the scattering vectors would be negligible. Both effects are not considered in classical approaches, like Williamson-Hall or Warren-Averbach methods, which analyze several (hkl) families simultaneously.

With the SLM analysis, grain size and microstrain can be calculated for each (hkl) reflection individually. The SLM separates size broadening

(Lorentzian peak shape) and strain broadening (Gaussian peak shape) via the ratio of FWHM and IBR for a single peak [Langford, 1978]. This is exemplarily shown in Fig. 3.9 with the (220) reflection of a tensile tested Pd thin film. In the investigated strain range, nearly all increase of IBR is of Gaussian shape, and corresponds to an increase in microstrain, while the Lorentzian share, respectively the grain size, is fairly constant.

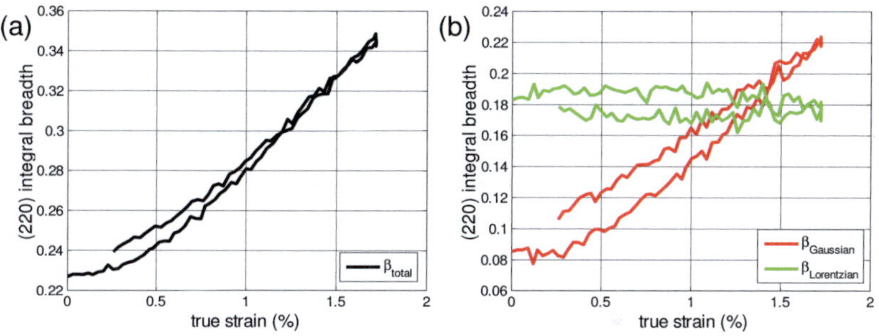

Figure 3.9.: The separation of (a) the IBR into (b) Lorentzian and Gaussian shares [Langford, 1978]. The example shows the separation of the (220) IBR of a Pd thin film tensile tested at the MS04 beamline of the SLS.

Instrumental broadening, determined with a standard powder reference (Y_2O_3 and CeO_2) is used for the correction of the recorded data [de Keijser et al., 1982]. The (hkl)-specific grain size D_{hkl} (more accurately: the coherent scattering domain size) and microstrain e_{hkl} are calculated according to

$$D_{hkl} = \frac{\lambda}{\beta_L^f \cos(\theta_{hkl})} \tag{3.5}$$

$$e_{hkl} = \frac{\beta_G^f}{4 \tan(\theta_{hkl})} \tag{3.6}$$

where λ is the monochromatic wavelength of the X-ray beam, β_L^f and β_G^f the Lorentzian and Gaussian broadening shares, corrected for instrumental broadening, and θ_{hkl} the Bragg angle. In the following, when microstrain is discussed, the root-mean-square microstrain, $<\varepsilon>$, is used, which is calculated from e_{hkl} according to

$$\langle \varepsilon_{hkl} \rangle = \langle \varepsilon_{0,hkl}^2 \rangle^{1/2} = \left(\frac{2}{\pi} \right)^{1/2} e_{hkl} \approx 0.7979\, e_{hkl} \qquad (3.7)$$

It is known that the SLM is not accurate with respect to the absolute value of D, as different values are obtained using different (hkl) reflections. In Fig. 3.10(a) the results of the SLM analysis for the (111), (200), and (220) reflections of a NC Ni bulk sample are shown. Indeed, different grain sizes ranging from 21 nm (220) to 30 nm (111) are determined. However, the initial grain shape is independent of the used reflection and indicates equiaxial grain shape in all cases. On the other hand, Fig. 3.10(b) shows the result for an exemplary Williamson-Hall analysis using the (111), (200), (220), (311), and (222) reflections for $\phi = 90°$. Similar to results in Ref. [Brandstetter et al., 2008] pronounced non-linear behavior resulting from elastic anisotropy and elastic grain interaction is observed. Similar to the SLM analysis, the derived grain size strongly depends on the selection of considered (hkl) reflections indicated by two exemplary linear regressions. The corresponding intercepts yield grain sizes in the range of 22 nm to 32 nm which is similar and by no means better than the results obtained by the SLM. Based on this comparison and the limitations of multiple peak approaches discussed above, the SLM was chosen to analyze the *in situ* diffraction data. In fact, the analysis and interpretation of the *in situ* diffraction data throughout this thesis rely only on relative changes of measured and derived parameters and is therefore independent of absolute values. The absolute grain size values are measured by complementary Automated Crystal Orientation and phase mapping (ACOM) measurements in conjunction with TEM (see Appendix A). This precise, independent, and direct microscopy investigation yields information on absolute grain size and orientation with nanometer spatial resolution. Thereby, it helps to prove and calibrate the grain size obtained by XRD. The ACOM data analysis was developed in close collaboration with Aaron Kobler and Dr. Christian Kübel (INT, Karlsruhe Institute of Technology), who carried out the TEM investigations.

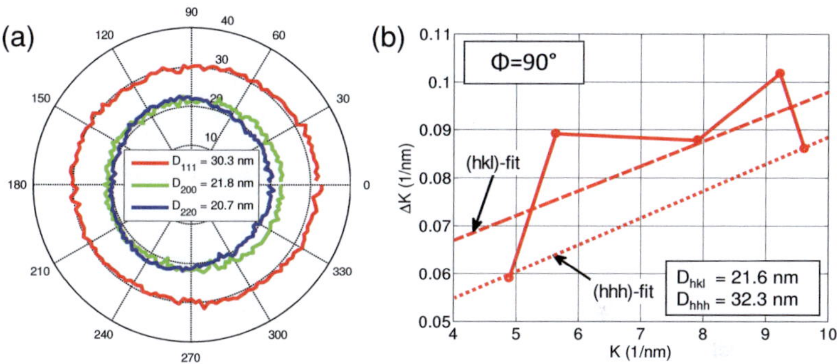

Figure 3.10.: Comparison of SLM and WH peak broadening analysis. (a) Initial in-plane grain size obtained by the SLM for different (hkl) planes for NC Ni. All planes yield an equiaxed grain shape, however absolute values vary from 21 nm to 30 nm. (b) Similar to the SLM, the WH analysis yields different absolute values for the grain size in the range of 22 nm to 32 nm depending on the considered (hkl) reflections. This was also found in Ref. [Brandstetter et al., 2008]. During deformation the WH analysis cannot be applied, since the different (hkl) planes experience different strain states.

3.5. Sample Preparation

The sample preparation as well as elementary initial characterization were carried out by cooperation partners. PED Ni was fabricated by Kerstin Schüler (AG Vehoff, Universität des Saarlandes), IGC PdAu alloys by the group of Prof. Rainer Birringer (Universität des Saarlandes), and Pd and PdAu thin films by Anna Castrup (INT, Karlsruhe Institute of Technology). The close collaborations ensured mutual feedback and improvement of the processes and sample quality.

3.5.1. Electrodeposited NC Nickel

Bulk Ni plates with a rectangular dimension of $40\,mm \times 70\,mm$ and a thickness between $1.5\,mm$ and $3\,mm$ are produced by pulsed electrodeposition (PED) from a Ni sulfamate electrolyte [Natter et al., 1998]. During this process, the microstructure of the deposit is controlled by the following deposition parameters: on-time, off-time, current density of the pulse function, and the concentration of organic grain refiners [Choo et al., 1995; El-Sherik et al., 1997]. Severe texture formation is suppressed by adding butynediol to the electrolyte. In addition, grain size is also controlled by the selected current density. Its increase promotes the formation of new nuclei rather than the growth of prevailing crystallites and thus also helps to suppress columnar growth [Kaischew, 1967; Moti et al., 2008].

The Ni samples, tested in this thesis, were prepared employing PED with a cathodic current density of $45\,mA/cm^2$ and setting a pulse length of $5\,ms$ followed by a $10\,ms$ off-time. The electrolyte was based on nickel sulfamate ($595\,ml/l$) with additives of nickel chloride hexahydrate ($5\,g/l$) and sodium lauryl sulfate ($0.2\,g/l$) [Hadian and Gabe, 1999]. Boric acid ($35\,g/l$) was used to buffer the pH-value. Furthermore, butynediol ($0.02\,g/l$) and saccharin ($0.4\,g/l$) were added to refine the grain size.

Regarding sample purity, an oxygen content of 4.5at% was detected using EDX, while heavy element impurities were beyond the detection limit of 0.5-1 at%. Particularly, no sulfur or chloride impurities from the electrolytic bath were detected.

The initial grain size of $D \approx 30\,\mathrm{nm}$ determined during the *in situ* synchrotron experiment was confirmed by an independent WH analysis in a laboratory diffractometer applying classical Bragg-Brentano geometry.

In contrast, the Ni from literature was mostly produced by Integran Technologies Inc. (Toronto, Canada), e.g. [Budrovic et al., 2004; Cheng et al., 2009; Wang et al., 2010]. When this material was no longer available, Ni-Fe alloys were used instead [Fan et al., 2006a,b; Li et al., 2008; Fan et al., 2009]. This is problematic since (i) the idea of independent proof of studies on pure Ni is violated and (ii) it is expected that the mechanical behavior of Ni-Fe alloys clearly differs from the behavior of pure Ni, especially with regards to the frequently discussed interfacial deformation modes. On the other hand, fabrication, microstructural characterization, and mechanical testing of self-fabricated material within a close collaboration allows to establish a self-consistent microstructure-property relationship.

Compression samples ($\approx 1 \times 1 \times 0.6\,\mathrm{mm}^3$) were cut by spark erosion from the Ni plates. In order to obtain a sample undergoing localized shear-dominated deformation, while externally a compressive load is applied, slits at $45°$ with respect to the loading direction were machined by spark erosion on the front and back side of a bulk sample ($\approx 3 \times 1 \times 0.6\,\mathrm{mm}^3$), as described in Ref. [Ames et al., 2010]. The width and height of the slit is $\approx 100\,\mu\mathrm{m}$. Beyond the sample preparation described in Ref. [Ames et al., 2010], the apex angle of the slit was increased in order to allow the under 2θ diffracted beam to exit the slit without penetrating the bulk material.

3.5.2. Inert-Gas Condensed NC PdAu Alloys

NC PdAu samples with different alloy compositions were prepared by inert-gas condensation (IGC). Small particles are condensed in an inert-gas atmosphere and subsequently consolidated at a pressure of 1.8 GPa to obtain disc-shaped samples with a diameter of 8 mm and a thickness of 0.5 mm to 1 mm [Birringer, 1989]. From these discs compression samples were prepared by spark erosion. The major advantage of the IGC method is that bulk samples can be fabricated with initially texture-free microstructures, and perfectly globular grains with D as small as ≈ 10 nm. As the microstructure of pure IGC Pd is not stable at room temperature [Ames et al., 2008], no mechanical tests are conducted on pure IGC Pd, because it would not have been possible to differentiate between thermally- and stress-induced changes of the microstructure. The nominal compositions of the investigated alloys range from 10 at% to 70 at% Au, in the continuously miscible PdAu alloy system. Relative densities of $> 95\%$ were determined using the Archimedes method, with a slight tendendy to less porosity for higher Au contents.

Prior to *in situ* mechanical testing, grain size and microstrain, as well as the exact alloy composition were characterized by classical WH analysis employing conventional XRD and Energy-Dispersive X-ray spectroscopy (EDX), respectively. The results are summarized in Table 3.2.

Table 3.2.: Summary of initial EDX and XRD characterization of IGC PdAu alloys.

	EDX		XRD	
Sample	Pd (at%)	Au (at%)	D (nm)	$<\varepsilon>$ (-)
Pd-10Au	88	12	9	0.005
Pd-30Au	66	34	7	0.005
Pd-50Au	43	57	9	0.006
Pd-70Au	23	77	10	0.005

3.5.3. Magnetron Sputtering of NC Pd and PdAu Thin Films

NC thin films with a typical thickness of $\approx 1\,\mu m$ consisting of pure Pd and various PdAu alloys were fabricated by RF magnetron sputtering. The films were deposited to flexible polyimide substrate (Kapton E, DuPont) with a thickness of 50 μm. Two inch targets with purities of 99.95% for Pd and 99.99% for Au were used. A detailed study on the influence of gas pressure and sputtering parameters on the microstructure of pure Pd and PdAu alloys is given in Ref. [Castrup et al., 2011]. Pure Pd was fabricated with a sputter power of 60 W and alloys with varying sputter powers of both targets. A summary of the sputter conditions is given in Table 3.3.

Table 3.3.: Summary of sputter parameters for pure Pd and PdAu thin films.

Power Pd (W)	60	60	60	60	60	30
Power Au (W)	0	15	20	30	60	60
Ar pressure (10^{-3} mbar)	5.0	5.0	5.0	5.5	5.5	5.0
EDX Au content (at%)	0	12	19	29	53	72

The base pressure was always less than 2×10^{-8} mbar. Pure Pd was fabricated by an interruptive process, where sputtering was always interrupted after 8 nm of film deposition using a fast rotational shutter, in order to avoid columnar grains. The alloys were prepared by continuous co-sputtering. During fabrication, the substrate holder was rotated steadily in order to obtain a homogeneous microstructure and a uniform alloy composition.

Prior to deposition, the substrates underwent an extensive cleaning procedure to achieve optimal conditions for good adhesion. The procedure included the steps of ultrasonic cleaning in an acetone bath, baking on a hot plate and finally baking inside the ultra high vacuum of the sputter chamber. The detailed procedure of substrate preparation is given in Ref. [Lohmiller et al., 2010; Lohmiller, 2009].

4. The Deformation Behavior of NC Bulk Metals and Alloys

4.1. Introduction

In the following chapter, experiments on different NC bulk metals and alloys are presented and discussed. In order to probe the bulk samples with typical dimensions in the mm range a high energetic X-ray beam is required. Therefore, all experiments were carried out at the High Energy Microdiffraction (HEMD) endstation of beamline ID15A at the ESRF providing an energy of $E_{Xray} = 69.7\,\mathrm{keV}$. The high penetration depth permits experiments in transmission geometry and thus provides excellent statistics. Using fast (up to 2 patterns per second) and large area detectors allows to collect up to 1000 diffraction patterns during one mechanical test. Each pattern comprises several complete Debye-Scherrer rings and thus the evolution of peak shape and texture can be monitored with high accuracy and time resolution. Particularly, texture analysis enables to differentiate between dislocation- and GB-mediated deformation mechanisms [Fan et al., 2006a; Ma, 2004; Markmann et al., 2003], since GB sliding and grain rotation lead to a randomized texture, while grain rotation based on dislocation plasticity promotes texture formation. For details on the experimental setup, the reader is referred to section 3.2.1.

Compression experiments allow to investigate the hitherto unexplored macroplastic regime ($\varepsilon > 10\%$) in detail, while the majority of *in situ* mechanical testing experiments has been conducted in tensile mode ([Budrovic et al., 2004; Cheng et al., 2009; Wang et al., 2010; Fan et al.,

2006b]), where the observation of fracture after a few percent of plastic strain is common evidence. Also the influence of other than uniaxial loading conditions on the materials response of NC metals has been mostly neglected but is indeed of scientific and technological interest, particularly the effect of shear-dominated deformation. Both antipodal deformation modes, namely dislocation- and GB-mediated plasticity should be affected: Dislocations require shear stresses to be activated. However, the role of dislocation-mediated deformation in NC metals is generally under heavy debate. On the other hand, it was recently shown, that GBs preferentially migrate under shear stresses, yielding to overall grain growth and preservation of the initial grain shape [Rupert et al., 2009]. Therefore, also shear-compression experiments were conducted on miniaturized shear-compression specimens (SCS) [Ames et al., 2010].

NC Ni ($D \approx 30\,\mathrm{nm}$) was used as a reference material for the development of the experimental setup and the data analysis. Besides common compression experiments, the applicability of this methodology (comprising high energy and microfocus of the X-ray beam) to *in situ* shear-compression experiments on miniaturized SCS was assessed with NC Ni. The electrodeposited samples were self-fabricated in close collaboration by Kerstin Schüler (Universität des Saarlandes) in order to address the issues of commercially available products discussed in section 3.5.1. A careful 3D characterization of the relevant microstructural parameters is required to establish a microstructure-property relationship. Therefore, *in situ* XRD is combined with accompanying high resolution Automated Crystal Orientation and phase Mapping (ACOM) analysis using TEM.

As a further aspect, alloying effects on the continuous miscible PdAu alloy system were investigated. Solute atoms are expected to pin GBs and hence reduce GB mobility, which should have a fundamental influence on the stability of NC alloys [Koch et al., 2008] and the contributions of

GB-mediated deformation mechanisms. Furthermore, the intrinsic stacking fault energy decreases in the alloy from pure Pd to pure Au by a factor > 4 [Schaefer et al., 2011], which should also have a significant impact on the deformation behavior, particularly on the occurrence and activity of partial dislocations. Sample preparation by IGC (see section 3.5.2) enables to obtain texture-free compression samples with a grain size of $D \approx 10\,\text{nm}$, which is exactly in that regime, where the Hall-Petch breakdown has been identified experimentally [Trelewicz and Schuh, 2007] and by modeling [Schiotz and Jacobsen, 2003]. Therefore it is expected, that, in addition to alloying effects, this peculiar grain size range intensifies the interaction of different deformation modes, especially the competing dislocation- and GB-mediated mechanisms.

The typical strain rates of the experiments were in the order of $\dot{\varepsilon} = 10^{-4}\,\text{s}^{-1}$. The setup is described in detail in section 3.2.1, sample preparation in section 3.5, and data analysis in section 3.3.

4.2. Results

4.2.1. Ni Compression Testing

Before the *in situ* mechanical tests on NC Ni samples are presented, it is noted that a detailed characterization of the initial 3D microstructure by ACOM/TEM, revealed the existence of columnar grains which are preferentially oriented along the growth direction and which exhibit an intragrain subdomain structure. For the first series of Ni compression tests, presented in the following, the orientation of sample (respectively columns), X-ray beam, and loading direction are shown in the schematic of Fig. 4.1(a): Load is applied perpendicular, whereas the incoming X-rays are oriented parallel to the growth direction. Consequently, as a result of the high energy of the X-rays ($E_{Xray} = 69.7\,\text{keV}$) the scattering vector is oriented almost perpendicular to the incoming beam and therefore the diffraction data, recorded on a two dimensional detector in transmission geometry, yield microstructural information within the cross-sectional planes (xy) of the columnar grains. In Fig. 4.1(b) and (c), the results of the 3D ACOM/TEM analysis of the initial 3D microstructure of NC Ni prepared by electrodeposition are displayed. The grains have an equiaxed grain shape with $D = 30\,nm$ in the xy-plane (plane normal parallel to growth direction) and elongated grains along the growth direction (z-direction).

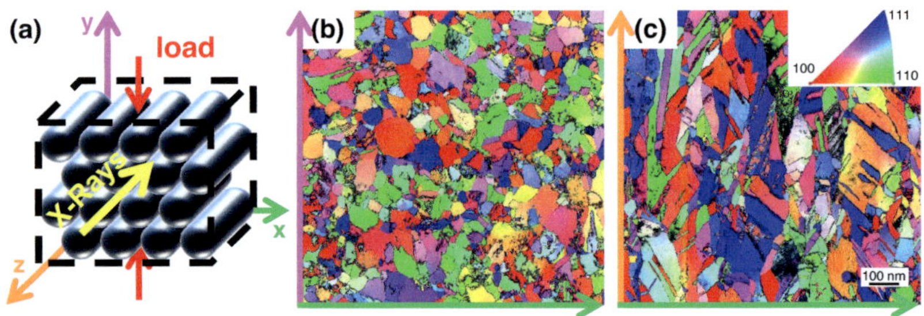

Figure 4.1.: Microstructure of PED NC Ni. (a) Schematic of the sample shows the slightly elongated grains along growth direction. For the first series of compression experiments, the X-ray beam is oriented parallel and the load direction perpendicular to the growth direction. Orientation maps obtained by ACOM/TEM are shown for (b) the xy-plane and (c) the xz-plane (see schematic).

A representative compressive stress-strain curve of NC Ni is displayed in Fig. 4.2(a). It is straightforward to subdivide the stress-strain curve into three regimes: A linear elastic regime (I), deviation from linear elasticity marking the onset of microplasticity (II), and macroplasticity (III) characterized by an almost constant strain hardening ($d\sigma/d\varepsilon \approx const.$). Applying load to the sample results in a concomitant shift of intensity maxima in radial direction (2θ) and changes of shape of the intensity profiles (Bragg-peak profiles) of constructively interfering (hkl) lattice planes as a function of the in-plane (azimuthal) angle ϕ. From peak shifts the evolution of (hkl) elastic lattice strains along the loading direction ($\phi = 90°$) has been derived and is shown in Fig. 4.2(b). The most compliant direction perpendicular to the (200) planes show highest elastic lattice strains while stiffer directions carry less strain. The pronounced differences become visible as a consequence of the very high overall lattice strains inherent to NC metals.

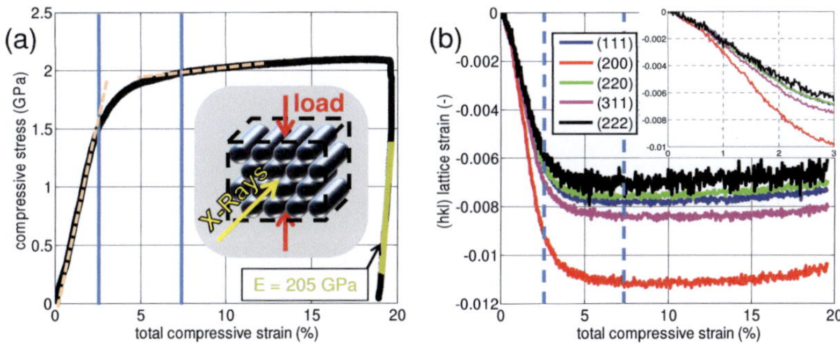

Figure 4.2.: *In situ* synchrotron compression experiment of NC Ni. (a) stress-strain curve. (b) (hkl) lattice strains over compressive strain ε_c for $\phi = 90°$, indicating that more compliant (hkl) families carry more elastic deformation, leading to strong lattice strain variations during deformation. The difference between the elastically equally compliant (111) and (222) direction is owed to the transmission geometry (cf. discussion in section 3.4 and Fig. 3.1(b)). The spreading of lattice strains starts immediately in the elastic regime, see inset. Overall high lattice strains are observed in comparison to CG materials reflecting the very high strength of the NC material.

Upon loading the progression of the integral intensity (INT) and the integral breadth (IBR) are related to texture formation and grain size as well as microstrain, respectively. In Fig. 4.3(a) and (b), the ϕ-dependent evolution of the INT and IBR of the (111) peak is shown. Data are displayed in polar plots for different total compressive strains $\varepsilon_c = 0\%$ (initial), 2.5% (regime I to II), 7.3% (regime II to III), and 19.5% (max.

strain). Regarding the INT, the onset of a redistribution of intensity is observed at the crossover from regime (I) to regime (II) to then develop a six-fold symmetry with maxima in intensity at every $60°$ and reduced intensities in-between. This observation indicates in-plane texture formation in NC Ni similar to the compression texture of CG FCC materials [Barrett and Massalski, 1966; Gambin, 2001]. The redistribution of the INT further proceeds during regime (III). The IBR, shown in Fig. 4.3(b), starts increasing uniformly until the crossover to regime (III) is reached. With further loading the IBR decreases and develops only a shallow ϕ-dependence.

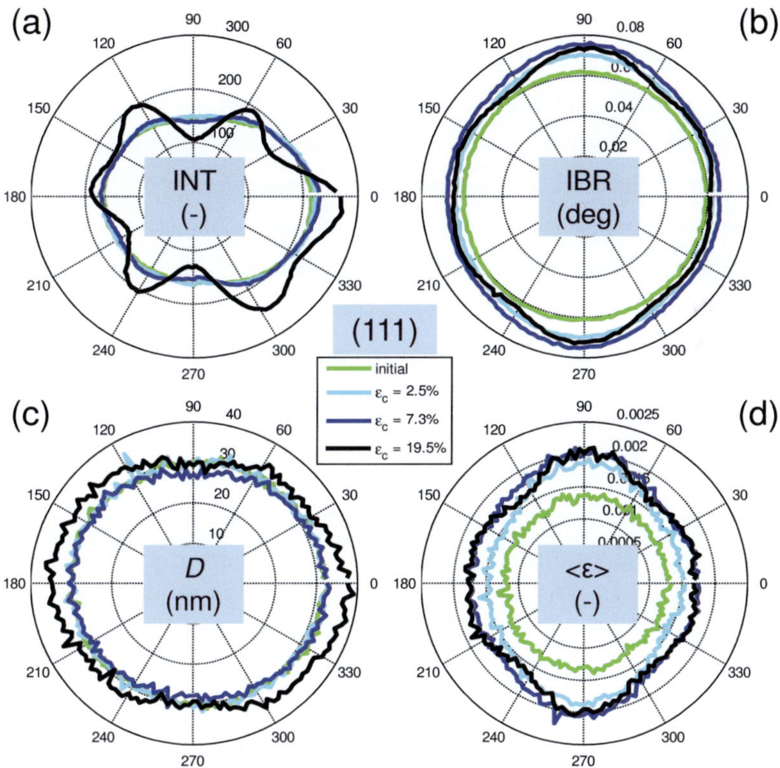

Figure 4.3.: Polar plots of (111) peak parameters and derived grain size and microstrain. (a) A 6-fold symmetry evolves with maxima at every $60°$ starting at $\phi = 0°$ from the initially uniform integral intensity (INT) distribution. This in-plane texture evolution demonstrates upcoming dislocation plasticity and is a typical compression texture known from CG FCC metals. (b) The initially uniform distributed integral breadth (IBR) increases in all ϕ directions until $\varepsilon_c = 7.3\%$ and slightly decreases afterwards, particularly in the same ϕ directions were INT maxima occur. (c) An elliptic grain shape develops from initially equiaxed grains ($D = 30\,\mathrm{nm}$) for $\varepsilon_c > 7.3\%$. Intriguingly, the grains grow in lateral direction but D remains constant in longitudinal compressive direction. (d) Initially uniformly distributed microstrain $\langle\varepsilon\rangle$ increases until $\varepsilon_c = 7.3\%$ and slightly decreases thereafter.

In order to extract grain size D and microstrain $<\varepsilon>$ values from the IBR, the SLM is utilized. The analysis is concentrated on the evolution of the (111) peak by putting special emphasis on high resolution of relative changes of D rather than on absolute values. The estimated error for the grain size evolution is ± 0.4 nm based on the scatter in Fig. 4.22. A detailed discussion on separation of size and strain and comparison with other techniques is given in section 3.4.2 (see also Fig. 3.10). Later, absolute values of grain size are independently verified by employing ACOM/TEM.

Inspection of Fig. 4.3(c) reflects initially equiaxed grains in the plane perpendicular to the X-ray beam and D remains constant up to the crossover to regime (III). With increasing strain, grain growth is observed in all directions except the loading direction, resulting in an elliptic grain shape. Note that D increases by $\approx 15\%$ in the lateral direction but does not decrease in the loading direction. The same behavior of constant D in longitudinal and $\Delta D/D_0 \approx 15\%$ in lateral direction is also obtained for other reflections.

It is shown in Fig. 4.3(d) that $<\varepsilon>$ is initially distributed uniformly along ϕ, to then strongly increase in regime (I) and approaching a maximum at the crossover to regime (III) in all ϕ directions. As a result, all additional peak broadening in regime (I) and (II) is related to increasing $<\varepsilon>$, whereas in regime (III) grain growth and concomitantly decreasing $<\varepsilon>$ lead to decreasing IBR, as seen in Fig. 4.3(b).

Texture evolution and grain growth extracted from the XRD data are confirmed by direct observation using ACOM in conjunction with TEM, ensuring that equivalent planes were probed. Details regarding the ACOM method and data evaluation are given in Appendix A. In the undeformed state (Fig. 4.4(a)), grains are equiaxed whereas after deformation grain growth is observed particularly in the horizontal tensile direction (Fig. 4.4(b) and (c)). Note that grains grow in this manner regardless of their orientation. In Fig. 4.4(d) the grain size of the undeformed and deformed state are compared quantitatively. Initially, grain size measures $D_{\phi=90°} = 29\,\text{nm}$ and $D_{\phi=0°} = 30\,\text{nm}$ in vertical and horizontal direction, respectively, and after deformation $D_{\phi=90°} = 31\,\text{nm}$ and $D_{\phi=0°} = 36\,\text{nm}$. Related to the ϕ direction of the maximum and minimum value of INT, correspondingly orientation maps are shown in Fig. 4.4(b) and (c). For $\phi = 90°$ the majority of grains is (110) oriented (green), whereas for $\phi = 60°$ the (111) (blue) and (100) (red) orientations dominate. In Fig. 4.4(e), the ϕ-dependent fraction of (111) oriented grains determined from inverse pole figures is displayed. Despite the limited volume examined by ACOM, a 6-fold symmetry is still observed after deformation, and so agrees with the *in situ* XRD results (Fig. 4.3(a)).

The grain sizes given in the former paragraph are number-weighted averages of the horizontal and vertical axes of all individual grain ellipses. The complete grain size distributions, in horizontal and vertical directions are given in Fig. 4.5. They are all mono-modal and can be described by log-normal distributions. The standard deviation of the log-normal distribution increased after deformation for the transversal direction, where grain growth is most pronounced. This has also been observed in Refs. [Rupert et al., 2009; Gianola et al., 2006]. However, neither bimodal grain size distributions nor anomalous grain growth is observed. A bimodal grain size distribution could also have been identified in the diffraction data by observing individual intensity hot spots of larger grains appearing superimposed to the homogeneous Debye-Scherrer rings of the smaller grains.

Since the careful initial 3D microstructural characterization revealed elongated grains and a $<111>$ texture along the growth direction (cf. Fig. 4.1), the possible influence of these microstructural features on the mechanical material response was investigated by additional compression tests with varying sample orientation relative to the loading direction, see Fig. 4.6. All investigated samples furnish the following evidence:

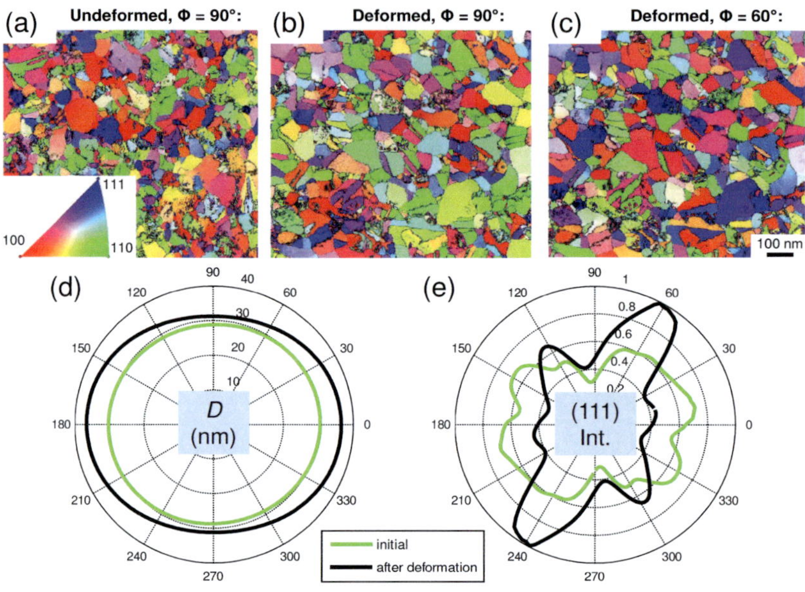

Figure 4.4.: Results from ACOM/TEM: Orientation maps of (a) the undeformed sample evaluated in $\phi = 90°$ and (b), (c) the deformed sample evaluated in $\phi = 90°$ and $\phi = 60°$, respectively. The color code for the orientation maps is shown as inset in (a). Polar plots with quantitative evaluation of (d) grain size from elliptic fits and (e) fractions of (111) oriented grains extracted from inverse pole figures evaluated in ϕ increments of $2°$. The results are consistent with the *in situ* XRD study.

Formation of (i) a similar 6-fold symmetry in intensity regardless of the initial in-plane intensity distribution, and (ii) an elliptical grain shape with apparently constant grain size in loading direction but growth of grains in lateral direction; likewise (iii) stress-strain curves are basically identical up to 10% strain (Fig. 4.6(a)-(c)) to then start slightly deviating from each other due to barreling and friction effects inherent to compression testing.

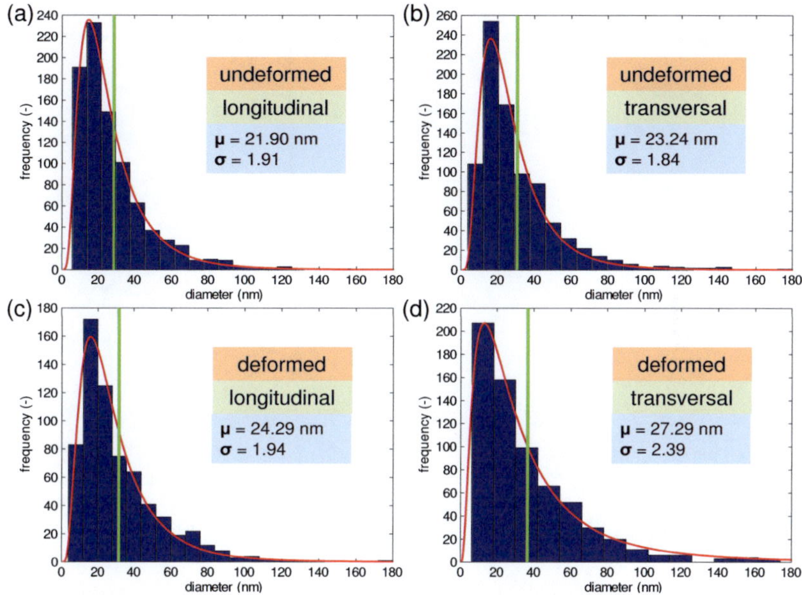

Figure 4.5.: Grain size distributions obtained by ACOM/TEM analysis. The grain sizes are evaluated in longitudinal and transversal direction for the undeformed and the deformed sample. (a) undeformed, longitudinal; (b) undeformed, transversal; (c) deformed, longitudinal; (d) deformed, transversal. All histograms can be fitted by a log-normal distribution (red lines). The green lines represent the corresponding number-weighted average values, which are used in Fig. 4.4(d).

4.2.2. Ni Shear-Compression Testing

Shear-compression specimens (SCSs) were prepared from the same PED NC Ni batch used for the compression samples in the former section. Therefore, the same microstructural feature of slightly elongated grains along the growth direction is present (cf. Fig. 4.1). For the shear-compression testing, the growth direction was oriented parallel to the incoming X-ray beam, similar to the main compression experiment schematically shown in Fig. 4.1(a).

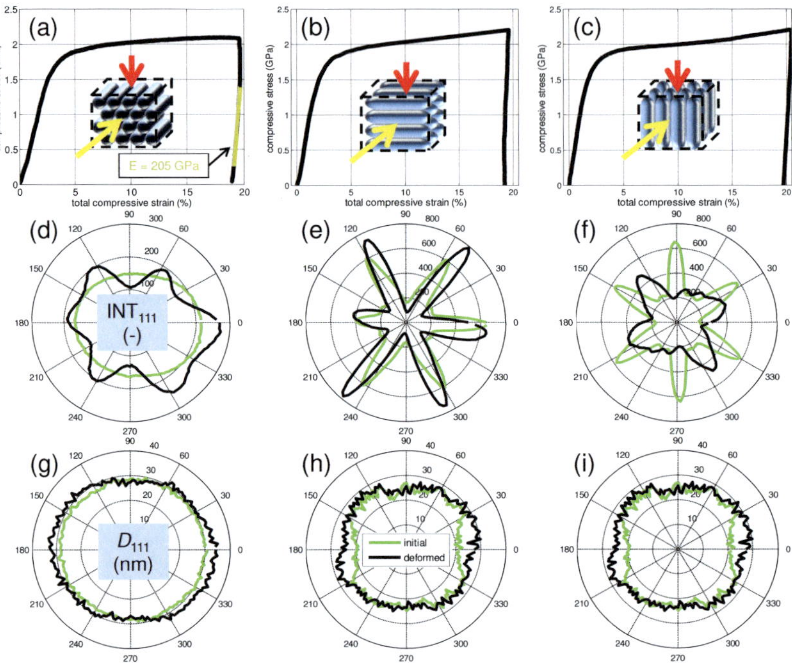

Figure 4.6.: Summary of the compression experiments with differently oriented samples: (a)-(c) show the stress-strain curves of the samples and the orientation of each sample is schematically illustrated in the insets. Comparison of the evolution of (111) integral intensity (d)-(f) and grain size and shape (g)-(i) for the initial (green) and the deformed (black) state.

The deformation state in the gauge section of a SCS is a complex three-dimensional stress and strain state. Therefore, direct calculation of the stress-strain curve from the measured load-displacement data is not possible. However, it was shown in Ref. [Dorogoy and Rittel, 2005], that an equivalent stress-strain curve can be obtained using the finite element method (FEM). The equivalent stress-strain curve for the NC Ni SCS obtained by FEM analysis is shown in Fig. 4.7(a). Details on the FEM simulation can be found in Appendix B. After a linear stress increase, the stress-strain curve traverses an extended microplastic regime and finally reaches a long macroplastic plateau.

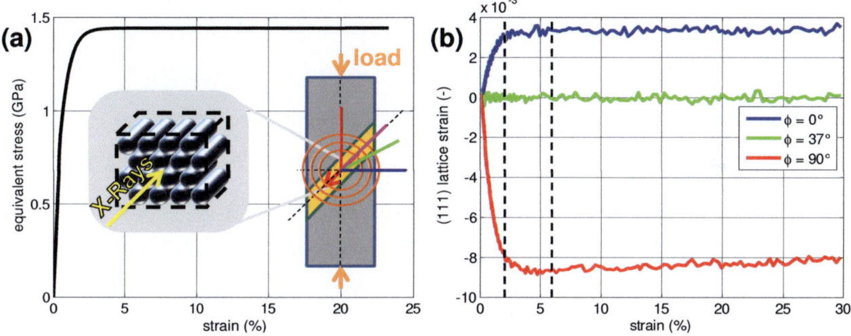

Figure 4.7.: (a) Equivalent stress-strain curve obtained by FEM analysis. The schematic in the inset shows the orientations of the growth direction and the X-ray beam. (b) Evolution of (111) lattice strains for different azimuthal directions ϕ over strain. The two vertical, dashed lines indicate the separation of different deformation regimes, analog to the pure compression test from the former section.

The area detector allows for monitoring the evolution of XRD peak parameters during loading along any azimuthal direction ϕ. Exemplarily, the (111) elastic lattice strain is plotted over strain ε for selected azimuthal directions, see Fig. 4.7(b). During initial loading, lattice strain starts to evolve: Compressive along the external loading direction ($\phi = 90°$) and tensile in the lateral direction ($\phi = 0°$), with a strain-free direction in-between, e.g. in the first quadrant at $\phi = 37°$. For $\varepsilon < 2\%$, the lattice strain behaves fairly linearly. After that, the lattice strain merges into saturation and at $\varepsilon \approx 6\%$, the absolute lattice strain reaches its maximum for each azimuthal direction. The sample fractures parallel to the slit after $\varepsilon = 30\%$. Until then, lattice strains remain constant in tensile directions, but decrease in compressive directions. The same behavior is observed for the other recorded (hkl) reflections, where only the plateau values differ, and are maximal for the (200) planes, which are elastically most compliant (e.g. for $\varepsilon = 6\%$: $\varepsilon_{200,90°} = $ -1.2% compared to $\varepsilon_{111,90°} = $ -0.85%).

Analogous to the pure compression sample (COMP) from section 4.2.1, the deformation behavior can be classified into three different regimes. For the SCS, this classification yields the following transitions: from regime (I) to (II), $\varepsilon = 2\%$ and from regime (II) to (III), $\varepsilon = 6\%$.

The difference of the in-plane deformation state of the SCS compared to the COMP tested under similar conditions becomes apparent when the elastic lattice strains as a function of the azimuthal direction ϕ are compared at a given state of deformation (see polar plots in Fig. 4.8).

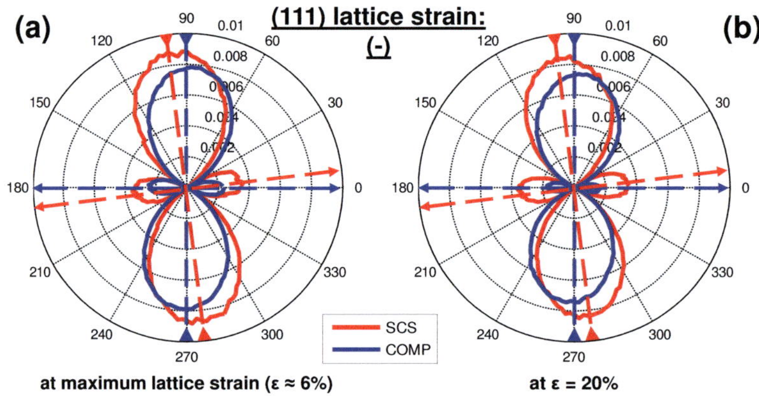

Figure 4.8.: Comparison of (111) lattice strains of the SCS and COMP at two different states of deformation: (a) When the maximal lattice strain values are reached ($\varepsilon \approx 6\%$) and (b) at $\varepsilon = 20\%$. Generally, the complete 2D strain state is rotated by $\phi_{offset} \approx 5.5\%$. Note that the inward pointing arrows indicate compressive strains and the outward pointing arrows tensile strains.

Generally, the complete two-dimensional strain state is rotated by $\phi_{offset} \approx 5.5°$ in the counter-clockwise direction. When considering the state of deformation at which the maximum lattice strains are reached in both samples (transition from regime (II) to (III), $\varepsilon \approx 6\%$), the maximum compressive lattice strain of the SCS is increased by $\approx 11\%$ compared to the COMP, and the lateral tensile lattice strain is even increased by $\approx 57\%$, see Fig. 4.8(a). Changes evolving during regime (III) are examined by comparison of Fig. 4.8(a) and (b): The compressive lattice strains slightly relax in both samples, as well as the lateral tensile lattice strain in the COMP. However, in the SCS, the tensile lattice strain remains constant during regime (III). Finally, at $\varepsilon = 20\%$, the lateral tensile strain of the SCS is twice as large as in the COMP. These (111) lattice strain values are summarized in Table 4.1.

Table 4.1.: Summary of (111) elastic lattice strains and the derived shear strains of SCS and COMP at two different deformation states: max. lattice strain (transition from regime (II) to (III), $\varepsilon \approx 6\%$) and at $\varepsilon = 20\%$. The shear strains were calculated from the principle strains according to the maximum shear stress theory from Tresca ($\varepsilon_\tau = (\varepsilon_t - \varepsilon_c)/2$) [Gross et al., 2005].

(111) elastic lattice strain (%)	max. lattice strain		
	comp.	tensile	shear
SCS	-0.88	0.36	0.62
COMP	-0.79	0.23	0.51
(111) elastic lattice strain (%)	$\varepsilon = 20\%$		
	comp.	tensile	shear
SCS	-0.84	0.36	0.60
COMP	-0.74	0.17	0.46

For the SCS, the normalized integral intensities $INT_{hkl,\phi}/INT_{hkl,\phi}^0$ of different (hkl) reflections are displayed in polar plots as a function of ϕ for the most relevant deformation states (Fig. 4.9(a)-(c)). The normalized intensities remain at unity up to a strain $\varepsilon = 2\%$. After that, the intensities redistribute along ϕ and a 6-fold symmetry starts to evolve for all monitored (hkl) reflections with maxima at every $60°$ similar to the COMP. This in-plane deformation texture particularly evolves during regime (III) ($\varepsilon >$ 6%). In the direction of maximum compressive load, (111) and (200) intensities deplete, while (220) becomes more intense. In the lateral tensile direction the opposite behavior is observed.

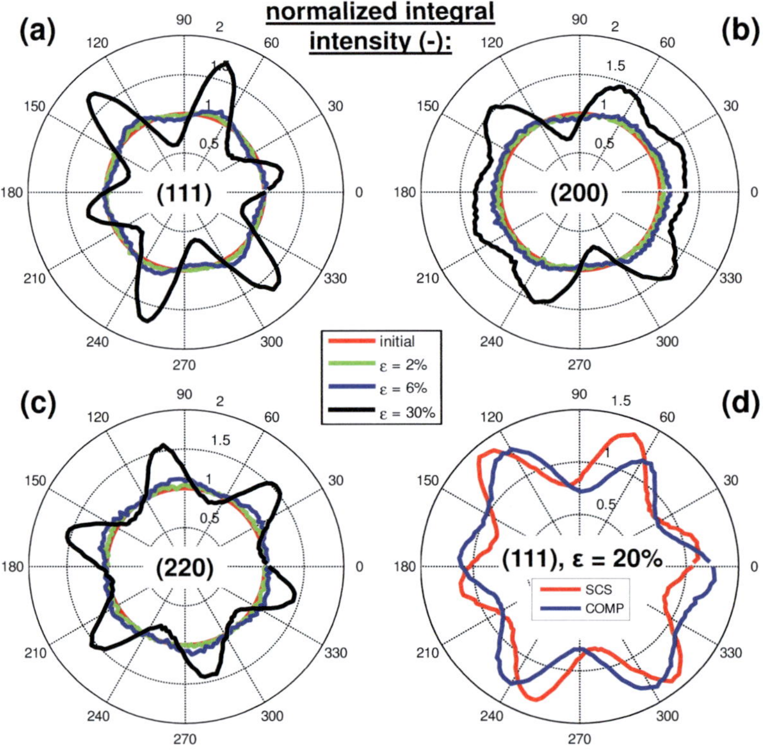

Figure 4.9.: The normalized integral intensities of the (a) (111), (b) (200) and (c) (220) reflections of the SCS are displayed in polar plots dependent on ϕ for different states of deformation. (d) The normalized (111) integral intensities of the SCS and COMP are compared at $\varepsilon = 20\%$.

In Fig. 4.9(d), the normalized integral (111) intensities of SCS and COMP at $\varepsilon = 20\%$ are compared. Two observations are pointed out: (i) The texture of the SCS is sharper, i.e. relating the values of the six INT maxima to the values of the six INT minima yields a texture ratio of 1.58 for the SCS and 1.44 for the COMP. (ii) The overall symmetry is rotated compared to the COMP.

In the next step, the in-plane rotations of lattice strain and INT are investigated in more detail. In Fig. 4.10, the evolution of (a) (111) lattice strain and (b) (111) INT is displayed exemplarily for a certain ϕ range as a function of total strain. In the colored maps the azimuthal angle ϕ is given on the abscissa and the strain ε on the ordinate. The color code indicates the amount of lattice strain and intensity, respectively. The white line, in the colored map for compressive lattice strain shown in the range of $60° < \phi < 120°$ as a function of ε, indicates the fitted ϕ positions of the maximum of compressive lattice strain (Fig. 4.10(a)). During loading, a constant deviation of $\phi_{offset} \approx 5.5°$ from the vertical loading direction ($\phi = 90°$) is observed, similarly to the in-plane rotation shown in Fig. 4.8. On the other hand, the INT evolution in the range of $100° < \phi < 160°$ and as a function of ε is shown in Fig. 4.10(b). From inspection of the colored map, a continuous rotation of the azimuthal peak maximum position is observed during loading. From the black line, representing the fitted peak maximum position, an in-plane rotation of the INT maximum position of $\Delta\phi \approx 8°$ is deduced for $\varepsilon = 30\%$.

In the schematic of Fig. 4.11(a), the initial azimuthal positions of the maxima of compressive and tensile (111) lattice strains are shown together with the maxima of (111) INT. The maximal positions of compressive ($\phi \approx 95.5°$ and $\phi \approx 275.5°$) and tensile ($\phi \approx 185.5°$ and $\phi \approx 365.5°$) lattice strains remain constant during deformation (Fig. 4.11(b)). However, all six (111) INT maxima rotate during deformation and an increasing ϕ angle for each maximum is noted (Fig. 4.11(c)). This means, that in addition

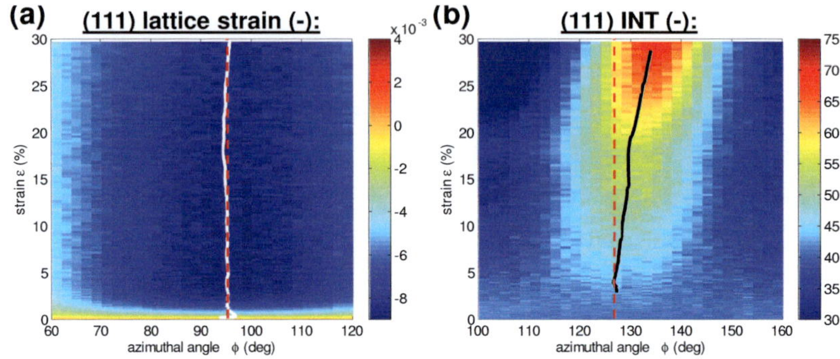

Figure 4.10.: Detailed analysis of the evolution of (111) lattice strain and INT as functions of ϕ and ε. (a) The ϕ position of the maximal (111) compressive strain around $\phi = 95°$ is constant over ε, while (b) a pronounced deformation-induced rotation to larger ϕ values is observed for the (111) INT maximum around $\phi = 130°$. The white and black line indicate the fitted ϕ positions for lattice strain and INT, respectively.

to the formation of the deformation texture, an overall directional rotation of the corresponding maxima and minima is observed. The fact that all INT maxima rotate in the same direction is attributed to the predefined deformation geometry of the SCS. Please note, that from the initially uniform INT distribution, which starts to redistribute for $\varepsilon > 2\%$, reliable fits are first obtained for strains $> \approx 4\%$, when some intensity variation along ϕ is present. Therefore, in Fig. 4.11(c), only data points from fits with an r-square value > 0.85 are shown. The average deformation-induced rotation of the six INT maxima is $\Delta\phi = 7°$.

Again, peak broadening analysis is used to derive microstructural parameters, namely grain size D and microstrain $<\varepsilon>$. But first, the integral peak breadth (IBR) is considered, which is later used for calculation of D and $<\varepsilon>$. The IBRs of the (111), (200) and (220) reflections are shown in the polar plots of Fig. 4.12(a)-(c) as a function of

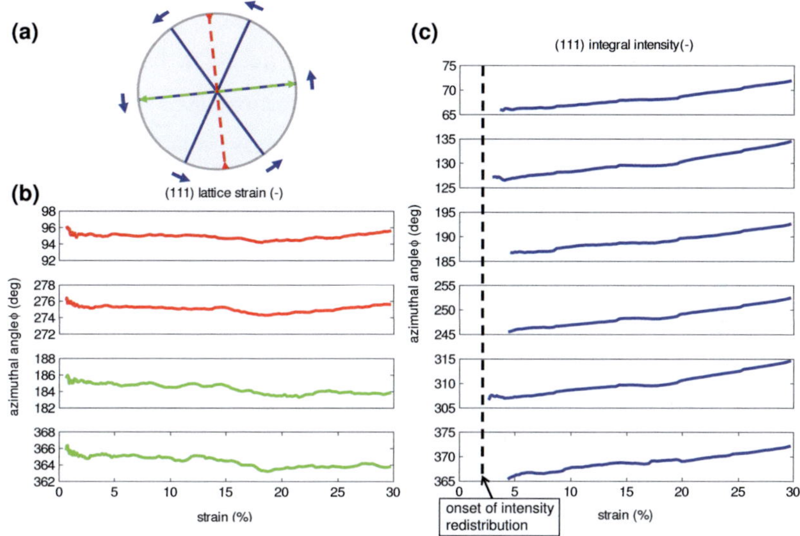

Figure 4.11.: (a) Schematic of the initial ϕ positions of maximal compressive and tensile (111) lattice strain, as well as of the (111) INT maxima. (b) and (c) Fitted ϕ positions of (111) lattice strain and (111) INT maxima, respectively. ϕ positions of all maxima of compressive and tensile strains remain constant during deformation, i.e. the 2D strain state is stable. All INT maxima rotate to larger ϕ values in a similar way. At the onset of texture formation ($\varepsilon = 2\%$), the INT maxima could not be fitted. Extrapolating the displayed trends back to 2% strain would yield an average overall rotation of $\Delta\phi = 9°$. The data represent the ϕ positions from Gaussian fits with a goodness of fit of r-square > 0.85.

ϕ for the same relevant deformation states as the intensities were examined before. At the beginning, the broadening is distributed uniformly along ϕ. After initial loading, the IBRs increase in all directions to their maximum values at $\varepsilon = 6\%$, which is also the state where lattice strains are maximal. At this state, the broadening is only marginally increased in the most compressive directions as compared to all other directions. For higher strains ($\varepsilon > 6\%$), the IBR starts to decrease. Comparison to the COMP at $\varepsilon = 20\%$ shows that under pure compressive load the 6-fold symmetry of (111) IBR is less pronounced compared to the SCS, but the IBR is rather dominated by maxima in the loading directions (see Fig. 4.12(d)). Likewise, the INT redistribution of the COMP is less pronounced (cf. Fig. 4.9(d) and lower texture ratio).

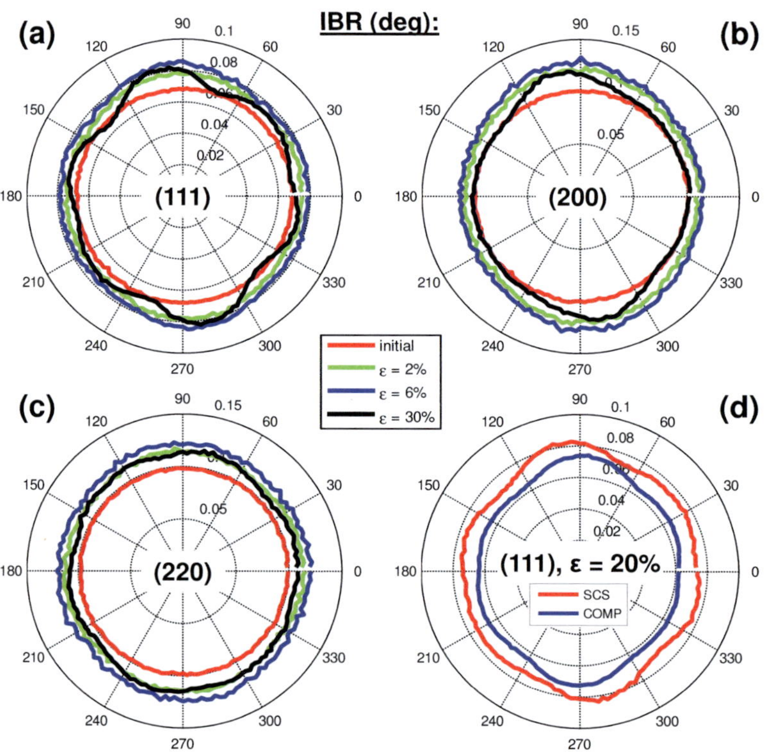

Figure 4.12.: The integral breadth of the (a) (111), (b) (200), and (c) (220) reflections of the SCS are displayed as polar plots for different states of deformation. In (d) the (111) IBRs of SCS and COMP are compared at $\varepsilon = 20\%$.

Particularly for the (111) reflection, the decrease of IBR occurs primarily in the directions, where the INT distribution shows its maxima. This results in a 6-fold symmetry for IBR, similar to that observed for the INT. Please note that maxima of IBR correspond to minima of INT and vice versa. The coincidence is displayed in Fig. 4.13 as a function of azimuthal angle ϕ.

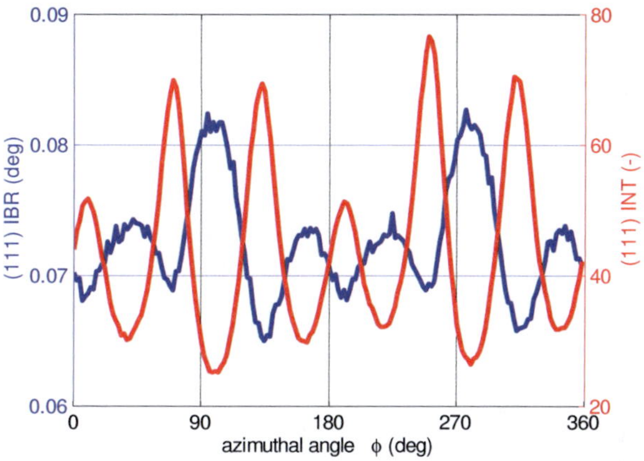

Figure 4.13.: Coincidence of the ϕ directions of (111) INT maxima and (111) IBR minima, and vice versa, at maximal deformation.

Applying the SLM to the broadening data yields grain size and microstrain values for each (hkl) reflection individually, as seen in Fig. 4.14. As it is well known, (hkl)-dependent absolute values are obtained from the SLM, as it was shown for the grain size of an undeformed Ni microstructure in Fig. 3.10(a). This is similarly observed for the SCS. However, it is surprising that also the grain shape yields (hkl)-dependent results during deformation. The grain shape analyzed from the broadening of the (111) planes reflects a six-fold symmetry with enhanced grain size in those ϕ directions, where the (111) IBR significantly decreases during regime (III), (cf. Fig. 4.12(a)). The (200) reflections yield a rather elliptic grain shape with enhanced grain size in the direction of lateral tensile strains, while the grain shape revealed by the (220) reflection does not reflect any change in grain shape or size. The microstrain, on the other hand, also reveals (hkl)-dependent absolute values, as expected, with the largest values for the (200) reflection. But the ϕ-dependent behavior of $<\varepsilon>$ is qualitatively fairly similar for the three reflections: Initially uniformly distributed, to then increase in all ϕ directions until a maximum is reached at $\varepsilon = 6\%$, with slightly stronger increases in the directions of maximum deformation. After the maximum, $<\varepsilon>$ decreases during regime (III).

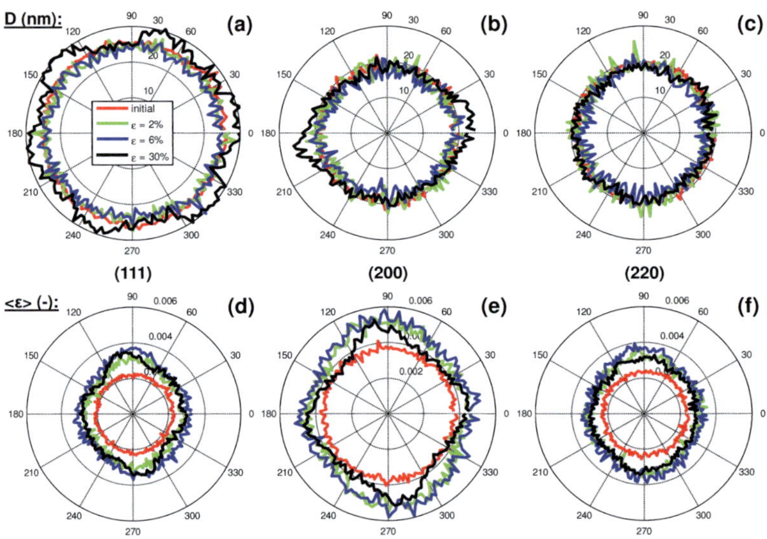

Figure 4.14.: Grain size (a)-(c) and microstrain (d)-(f) obtained from the SLM for (111), (200), and (220) presented as polar plots for relevant states of deformation.

Note that the decrease of $<\varepsilon>$ is dependent on the combination of (hkl) reflection and ϕ direction: In the direction of maximum compressive lattice strains ($\phi \approx 95.5°$ and $\phi \approx 275.5°$), (111) and (200) reflections exhibit only slight decreases while (220) exhibits pronounced decreases. Rotated to these ϕ directions by $\pm 30°$, the opposite behavior is observed with strongest decreases for (111) and (200) and lower decreases for (220). Comparing these observations to the INT redistributions shown in Fig. 4.9(a)-(c), it is seen that the strong decreases of $<\varepsilon>$ correspond to pronounced gains in INT and weak decreases of $<\varepsilon>$ to an INT depletion. Please note, that this observation is independent of the considered (hkl) reflection.

ACOM/TEM analysis is used to determine absolute grain size and grain orientations (see Fig. 4.15), serving as comparison for the *in situ* XRD results. For details on the method, the reader is referred to Appendix A. The orientation map (Fig. 4.15(a)) is recorded from a TEM lamella extracted directly at the fracture site of the SCS. Compared to the COMP sample (Fig. 4.4(b) and (c)), an enhanced amount of twins is noticeable. The twins primarily occur parallel to the slit of the SCS. In Fig. 4.15(b) and (c), polar plots are displayed with the quantitative analyses of grain size and grain orientation, respectively. The in-plane grain shape develops a distinct elliptic form during deformation compared to the initially equiaxial grain shape. Both, grain shape change and overall isotropic growth are observed. The grain size after deformation should yield an upper bound for grain growth, since the lamella was directly extracted at the fracture site, where grain growth and plastic deformation should be most pronounced. The in-plane distribution of grain orientation yields a similar relative redistribution behavior for all (hkl) families as it was also identified by XRD (cf. Fig. 4.9). Please note, that the XRD data cannot be directly laid on top of the ACOM data, since the exact angular position of the lamella cannot be exactly traced back. As it was mentioned, the lamella was extracted at the fracture site, which is not very plain, but rather rough. Therefore some offset angle between XRD and ACOM data is very likely.

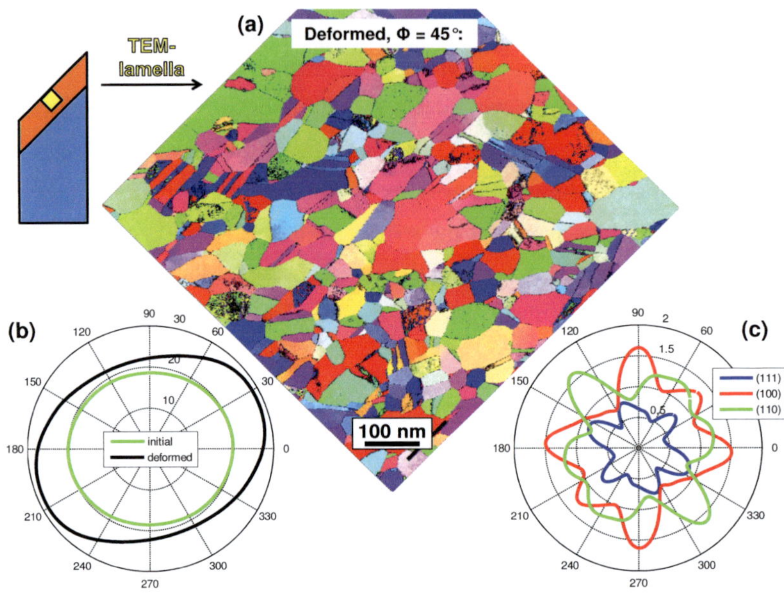

Figure 4.15.: ACOM/TEM analysis for NC Ni SCS. (a) Orientation map is displayed for the slit direction ($\phi = 45°$). The TEM lamella was extracted directly at the fracture site, see schematic. Polar plots with quantitative evaluation of (b) grain size from elliptic fits and (c) fractions of differently oriented grains extracted from inverse pole figures evaluated in ϕ increments of $1°$.

4.2.3. PdAu Compression Testing

In order to investigate the influence of alloying on the mechanical behavior of NC metals, different PdAu alloys prepared by IGC (see section 3.5.2) were tested under compressive loading. The nominal compositions ranged from Pd-10Au with 10 at% Au to Pd-70Au with 70 at% Au in steps of 20 at%.

The macroscopic compressive stress-strain curves of the tested alloys are displayed in Fig. 4.16. The Pd-70Au sample clearly exhibits the lowest absolute stress level. However, the strain hardening behavior is similar to that manifested by samples Pd-30Au and Pd-50Au. Pd-10Au does not show this pronounced increase of flow stress during plastic deformation, as a result of crack formation parallel to the compressive loading direction for $\varepsilon_c > 7\%$. This relatively brittle response of Pd-10Au was observed for several samples during *in situ* compression testing and was independently found in Ref. [Kurmanaeva et al., 2010].

In Fig. 4.17, the behavior of (111) lattice strain and peak asymmetry for the different alloys is shown. The evolution of the (111) elastic lattice strain in compressive direction ($\phi = 90°$) is shown in Fig. 4.17(a). Initially, linear behavior is observed for all alloys. However, for $\varepsilon_c > 7\%$ the curve of Pd-10Au deviates from the others, saturates and then decreases with further loading. Similar to the macroscopic stress in Fig. 4.16, the lattice strain is affected by crack formation, which inhibits a further build-up of lattice strain. For the other alloys, lattice strain increases continuously without saturation, even for high plastic strains. The ϕ-dependent lattice strain at $\varepsilon_c = 10\%$ is shown in the polar plot of Fig. 4.17(b). In the vertical direction maximum compressive strains are measured, and in the horizontal direction maximum tensile strains. The ratio of absolute longitudinal to transveral strains is ≈ 0.4 for all alloys, corresponding to the Poisson's ratio of 0.39 for Pd and 0.42 for Au [Beck, 1995]. Absolute lattice strain values are

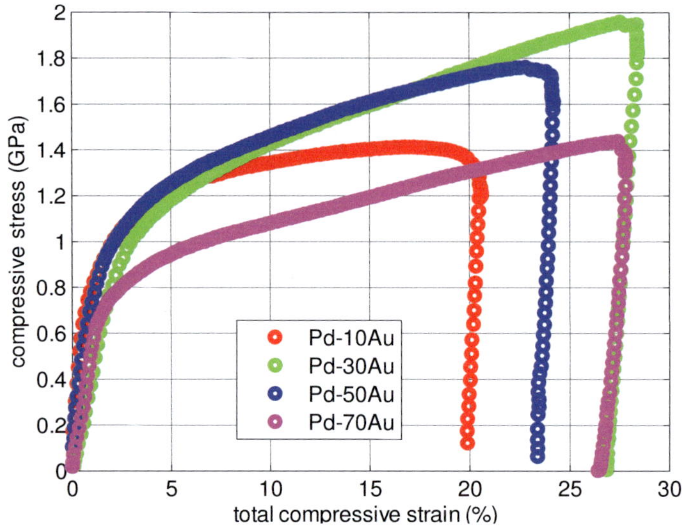

Figure 4.16.: Macroscopic compressive stress-strain curves of different PdAu alloys.

highest for the intermediate alloys, Pd-30Au and Pd-50Au and lowest for the already cracking Pd-10Au.

During loading a pronounced peak asymmetry evolves from initially almost symmetric peaks as shown in Fig. 4.17(c) and (d). In the ϕ directions of compressive strains (peak shift to larger 2θ values) a left-skewed asymmetry evolves ($A > 0$) and vice versa. Intriguingly, the trend of asymmetry vs. ε_c is similar to the lattice strain evolution: Initially, linear behavior is observed and for higher plastic strains, no saturation but still slightly increasing behavior is measured.

In the following, results on (hkl) peak intensities, their evolution over strain and redistribution along ϕ are presented. In Fig. 4.18, the ϕ-dependent evolution of the (111) integral intensity is considered. For all alloys, from the initially uniformly distributed INT (Fig. 4.18((a)), a six-fold symmetry evolves during deformation, shown at $\varepsilon_c = 20\%$ in

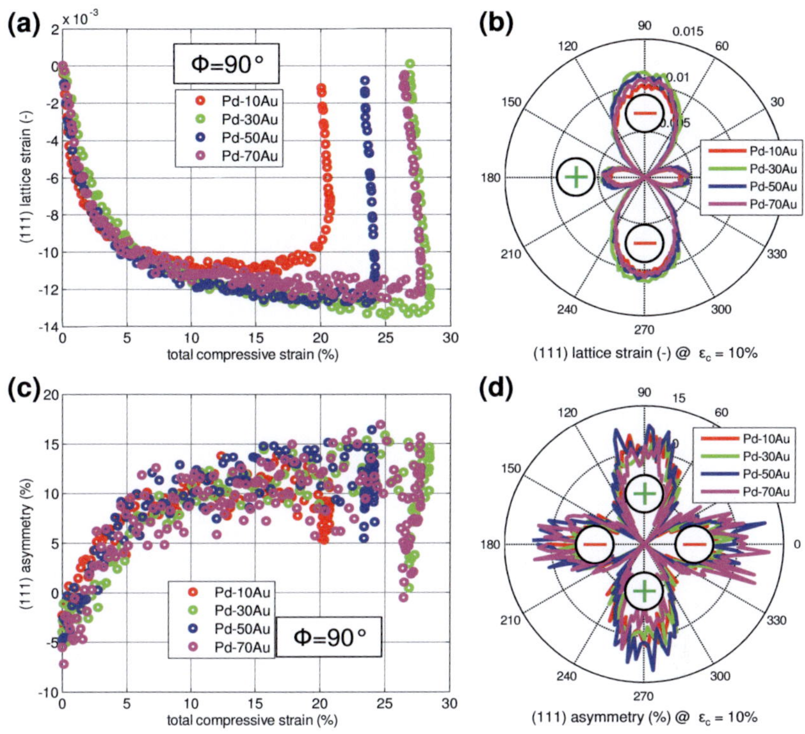

Figure 4.17.: Alloy-dependent (111) peak behavior: (a) lattice strain over ε_c for $\phi = 90°$, (b) ϕ-dependence of lattice strain at $\varepsilon_c = 10\%$, (c) asymmetry over ε_c for $\phi = 90°$, and (d) ϕ-dependence of asymmetry at $\varepsilon_c = 10\%$

Fig. 4.18(b). By inspection of the polar plot, it is seen that the sharpness of texture is composition-dependent, and increases with increasing Au content.

The absolute INT values in Fig. 4.18(a) and (b) are affected by exposure time, sample thickness, and the alloy-dependent mass absorption coefficient. In order to evaluate the texture formation without these effects and for better comparison, the (111) INT distribution at $\varepsilon_c = 20\%$ (Fig. 4.18(b)) is normalized to the initial value taken from Fig. 4.18(a) and

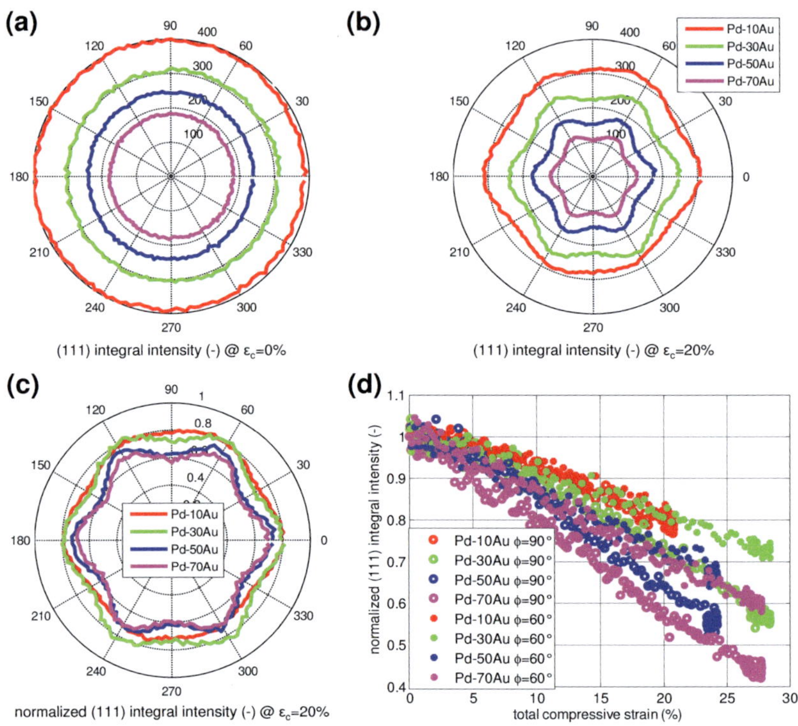

Figure 4.18.: Alloy dependence of (111) INT. (a) Initial, (b) after $\varepsilon_c = 20\%$, (c) after $\varepsilon_c = 20\%$ and normalized by the initial values, and (d) evolution of normalized INT over ε_c for $\phi = 90°$ and $\phi = 60°$.

plotted in Fig. 4.18(c). Furthermore, the evolution of the normalized INT as a function of ε_c is represented by two ϕ directions ($\phi = 90°$ exhibiting (111) minima and $\phi = 60°$ exhibiting (111) maxima) in Fig. 4.18(d). Interestingly, the INT globally decreases in all ϕ directions (all normalized INT values are less than 1), instead of increasing INT maxima and decreasing INT minima, with increasing ε_c. The decrease of INT is more pronounced for $\phi = 30°, 90°, 150°$, and so on, compared to ϕ directions thereto rotated by $30°$, forming the observed six-fold symmetry.

The overall decrease of INT will be discussed in detail in section 4.3.3 in terms of the ratio of background to peak intensity [Ungar et al., 2005].

The normalized INTs of (200) and (220) after $\varepsilon_c = 20\%$ are displayed in Fig. 4.19. In the unloaded state, the distributions of both reflections are uniform along ϕ, as also observed for (111), see Fig. 4.18(a). A six-fold symmetry evolves for the (200) planes, with directions of maxima and minima similar to (111), however less rotationally symmetrical. In fact, in the main compressive direction, $\phi = 90°$ and $\phi = 270°$, the decrease in intensity is most pronounced, meaning that one preferred orientation develops and strictly speaking, the six-fold symmetry reduces to a two-fold symmetry (i.e. mirror symmetry). The INT of the (220) develops a six-fold symmetry as well, but maxima and minima are rotated by $30°$ compared to (111) and (200), resulting in INT maxima at $\phi = 30°, 90°, 150°, \ldots$ and minima in-between. As also observed for the (200), the relative change of INT is most pronounced in the compressive direction.

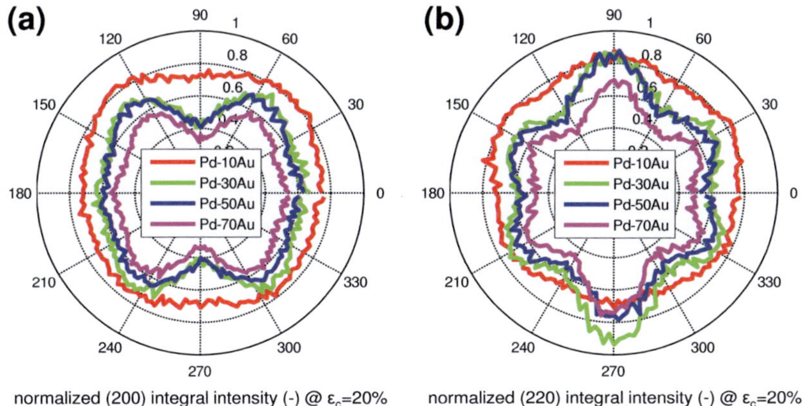

Figure 4.19.: Alloy dependence of normalized INTs of (a) (200) and (b) (220) after $\varepsilon_c = 20\%$.

In order to quantify the sharpness of texture, the averaged INT values of the six maxima are divided by the averaged INT values of the six minima, at $\varepsilon_c = 20\%$. The calculated values for (111), (200), and (220) are displayed in Table 4.2. For all reflections, stronger texture formation for higher Au contents is observed. Furthermore, the ratio of (200) to (111) texture also increases with increasing Au contents. Please note, that the normalization of INT does not affect the calculated texture ratios, since initially a uniform INT distribution is existent (cf. Fig. 4.18(a)).

Table 4.2.: Alloy- and (hkl)-dependence of the texture parameter evaluated at $\varepsilon_c = 20\%$.

Sample	(111)	(200)	(220)	(200)/(111)
Pd-10Au	1.11	1.17	1.12	1.05
Pd-30Au	1.18	1.45	1.23	1.23
Pd-50Au	1.23	1.53	1.29	1.24
Pd-70Au	1.24	1.56	1.45	1.26

Finally, peak broadening data, and resultant grain size and microstrain, calculated by the SLM, are presented. The alloy-dependent evolution of the (111) IBR over ε_c for $\phi = 90°$ is shown in Fig. 4.20(a). Initially, the intermediate alloy compositions exhibit highest IBR values, while the alloys closer to pure materials show reduced IBRs. The trends during deformation are qualitatively similar for all alloys: Initially, only slight increases in IBRs are measured, but at high plastic strains ($\varepsilon_c \approx > 8\%$) IBRs decrease with further loading. After unloading, the IBRs of all alloys are below their initial values. In Fig. 4.20(b), the ϕ-dependent behavior of the (111) IBR is displayed. The upper half of the polar plot shows an initially uniform distribution of IBR with lowest values for Pd-10Au and Pd-70Au and highest values for Pd-30Au, as also seen on the ordinate in Fig. 4.20(a). At $\varepsilon_c = 20\%$, shown in the lower half of the polar plot, the IBR distribution is elliptic for all alloys. In compressive direction, the

IBR decreased only slightly compared to the initial state. However, more pronounced reductions are observed in the lateral tensile directions.

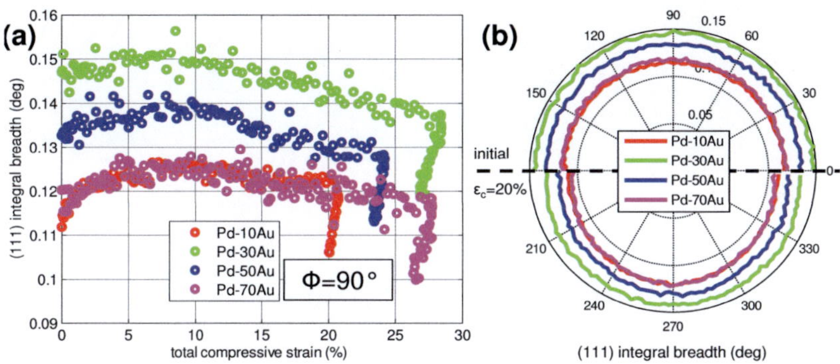

Figure 4.20.: Alloy dependence of (111) IBR. (a) Evolution of IBR over ε_c for $\phi = 90°$ and (b) ϕ-dependence of IBR: Upper half shows the initial distribution and the lower half the distribution at $\varepsilon_c = 20\%$.

As before, the SLM is applied in order to compute grain size D and microstrain $<\varepsilon>$. The results from the (111) broadening data are shown in Fig. 4.21. Initially, all alloys exhibit an equiaxed grain shape, with largest grain sizes for Pd-10Au ($D = 14$ nm) and smallest grain sizes for Pd-30Au ($D = 10$ nm). After $\varepsilon_c = 20\%$, all alloys underwent grain growth. However, the grain shapes remain fairly equiaxed. The relative grain size increase, $< \Delta D > \phi / < D > \phi$, is measured by relating the ϕ-averaged ($0° < \phi < 360°$) grain size change at $\varepsilon_c = 20\%$, to the ϕ-averaged initial grain size. In contrast to the initial grain size, the relative grain size changes show an alloy dependence with stronger increases for higher Au contents, as displayed in Table 4.3.

The initially uniformly distributed $<\varepsilon>$ is highest for intermediate alloy compositions and lowest for Pd-70Au. At $\varepsilon_c = 20\%$, for all alloys, the $<\varepsilon>$ is elliptic in shape with highest values in the compression directions.

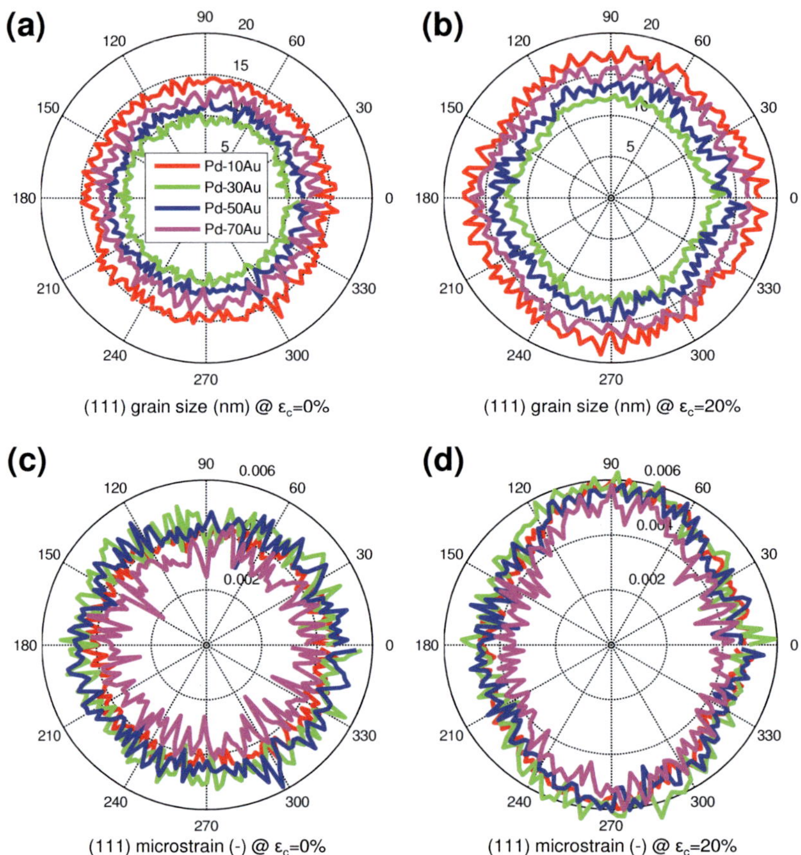

Figure 4.21.: Results from SLM evaluated for (111). Grain size D (nm) (a) in the initial state and (b) after $\varepsilon_c = 20\%$ and microstrain $<\varepsilon>$ (c) initially and (d) after $\varepsilon_c = 20\%$.

In the lateral tensile direction the $<\varepsilon>$ values are similar to the values in the initial state.

Table 4.3.: Alloy-dependence of the relative grain size increase at $\varepsilon_c = 20\%$.

Sample	Pd-10Au	Pd-30Au	Pd-50Au	Pd-70Au
$\langle \Delta D \rangle_\phi / \langle D \rangle_\phi$ (%)	16	16	18	20

4.3. Discussion

Compression and shear-compression experiments on NC Ni and compression experiments on NC PdAu alloys were carried out using a high energy XRD *in situ* mechanical testing setup. The following discussion of the obtained results will be subdivided, examining different aspects individually. At first, the succession of different deformation mechanisms will be discussed based on the analysis of the results of Ni compression testing. Then, the influence of the loading condition on peak parameters and deformation behavior will be discussed, comparing pure compression with shear-compression experiments. In the third part, alloying effects on the deformation behavior of PdAu compression samples will be discussed and the behavior will be compared to pure Ni. Finally, the most important findings and insights will be summarized.

4.3.1. Succession of Deformation Mechanisms

The following discussion mainly focuses on the results of Ni compression testing. Many studies exist on NC Ni and thereby it can serve as a reference material for the verification of the developed setup and data analysis. In order to gain more insight into the deformation mechanisms, first it is scrutinized whether the distinct regimes of deformation seen in the macroscopic stress-strain curves (Fig. 4.2(a)) are also reflected in the microstructural data derived from *in situ* XRD (Fig. 4.22(a)-(b)). Fig. 4.22(a) displays the normalized INT of the (111) peak for the loading direction ($\phi = 90°$) and for $\phi = 60°$, which are directions of an intensity minimum and maximum related to texture formation, respectively (Fig.

4.3(a)). Obviously, there is no intensity redistribution up to 3% compressive strain, thus implying absence of dislocation glide. Since non-linear elastic behavior cannot be ruled out, the 3% strain value manifests an upper bound of elastic material behavior. With further increasing load, dislocation activity becomes gradually more important entailing pronounced intensity redistribution beyond 10% strain. A prominent feature of Fig. 4.22(b) is the correlation of the maxima in lattice strain and microstrain with the onset of grain growth along the tensile direction ($\phi = 0°$ and $\phi = 180°$). Interestingly, the strain value of 7.3% associated with the maxima favorably agrees with the onset of regime (III) extracted from the stress-strain curve. Overall, the footprint of the evolution of microstructural parameters independently reflects again three distinct deformation regimes.

Now the aim is to identify physical deformation mechanisms to explain the obtained overall data base. In regime (I), the initial loading is dominated by build-up of linear elastic lattice strain. With increasing load, the lattice strains reach high values of up to 1% and reveal a pronounced (hkl) dependence (Fig. 4.2(b)). Consequently, depending on orientation and size, a broad spectrum of elastic deformation develops among individual grains. Locally varying strain in conjunction with the implicitness of overall compatible deformation necessitates accommodation processes in or at GBs and triple junctions, e.g. atomic shuffling [Derlet et al., 2003b], and contributes to an increase in $<\varepsilon>$. Dislocation activity is excluded as a possible reason for the increase in $<\varepsilon>$ in regime (I) since build-up starts from the very beginning of the experiment, when stresses especially in the grain interior are rather low. In fact, sophisticated analysis of MD simulations [Markmann et al., 2010; Bachurin and Gumbsch, 2010] also show immediate increase of $<\varepsilon>$, even though dislocation activity is retarded to strains above 3.5%. Moreover, the recently observed shear softening of GBs in NC Pd [Grewer et al., 2011] supports the idea that GB-mediated shear and slip [Weissmueller et al., 2011] may carry a

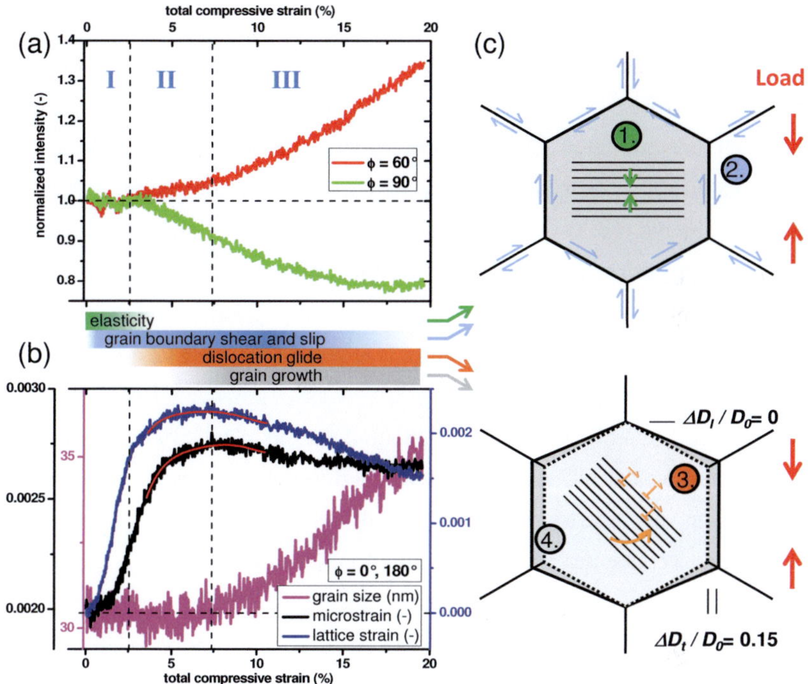

Figure 4.22.: (a), (b) Masterplots of the most relevant XRD peak parameters as a function of total compressive strain and (c) schematic of the identified microscopic deformation mechanisms. An explicit succession of different deformation modes can be derived: (I) Inhomogeneous elastic lattice straining and GB accommodation, (II) upcoming dislocation plasticity, inferred from texture evolution, and (III) onset of stress-driven GB migration.

dominant share of macroscopic deformation at stress levels too low to mobilize dislocations.

In regime (II), GB-mediated deformation proceeds and leads to increasingly inhomogeneous strain and stress in the NC aggregate until local stresses become high enough to activate already existing dislocation embryos within single grains [Bachurin and Gumbsch, 2010] or give rise to dislocation nucleation at individual GBs [Van Swygenhoven et al., 2004]. With ongoing deformation this will occur for an increasing number of grains, governing the transition from micro- to macroplasticity [Saada, 2005; Brandstetter et al., 2006]. Similar to a coarse-grained polycrystalline FCC material, dislocation plasticity will lead to a deformation texture as indicated by the 6-fold symmetry in INT (Fig. 4.3(a)). A single dislocation traversing a grain can carry strain of the order of b/D, where b is the Burger's vector ($b/D \approx 0.8\%$ for Ni with $D \approx 30\,$nm) and hence also contributes to grain rotation ($\arctan(b/D) \approx 0.5°$) as well as a directional change in grain shape. As GB-mediated processes still carry a considerable share of the overall deformation and dislocation plasticity is inherently constrained by the NC microstructure (high activation stress, only single dislocations per grain) the texture evolution is much less pronounced compared to conventional coarse-grained FCC materials.

With increasing total compressive strain ($\varepsilon_c > 7.3\%$) the threshold for stress-driven GB migration (SDGBM) [Rupert et al., 2009; Cahn et al., 2006] is reached, which results in isotropic grain growth during regime (III). In addition, a decrease of $<\varepsilon>$ is observed. Since a source of microstrain in NC metals are compatibility stresses which give rise to a $<\varepsilon> \propto 1/D$ scaling [Ames et al., 2008], it seems plausible to interpret the decrease in $<\varepsilon>$ as being correlated to the increase in D. In fact, the analysis of grain growth in NC Pd yields a slope of $<\varepsilon>$ versus $1/D$ of 0.020 [Ames et al., 2008], which is in good agreement with the slope of 0.023 found for our data in regime (III).

The coexistence of SDGBM - that leads to isotropic grain growth - and dislocation plasticity - that results in grain elongation in tensile and shrinkage in compressive direction - manifests elliptical grain shape. The specific evolution of grain shape may thus allow to discriminate between dislocation plasticity and SDGBM. The almost constant D in longitudinal direction reflects a cancellation of both contributions, whereas they add up to elongation perpendicular to the loading direction (Fig. 4.22(c)). Assuming that the active glide systems are working predominantly under 45° with respect to the loading axis, shrinkage and elongation of the grains by dislocation plasticity is equal in magnitude.

From the experimental data for the grain size evolution, it was found that the initial grain size of the tested samples is $D_0 \approx 30\,\mathrm{nm}$ and the perpendicular elongation is about 4.5 nm, so entailing $2.25\,\mathrm{nm}/30\,\mathrm{nm} = 7.5\%$ of strain along the loading direction contributed by dislocation plasticity. This evidence implies that dislocations carry only about 40% of the overall 19% compressive plastic deformation of the sample. Consequently, SDGBM and GB-mediated plasticity should be responsible for the remaining 60% of deformation. Assuming an average coupling factor $\beta = 0.4$ [Cahn et al., 2006], SDGBM contributes about 3% strain when GBs migrate on average 2.25 nm, corresponding to about 15% of the overall deformation. Consequently, the remaining 45% of the overall deformation must have been accommodated by GB-mediated deformation processes. The corresponding equations used for the described approach are summarized in the following:

$$\Delta D_{disl} - \Delta D_{migr} = 0\,\text{nm} \qquad \text{(compressive direction)}$$

$$\Delta D_{disl} + \Delta D_{migr} = 4.5\,\text{nm} \qquad \text{(tensile direction)}$$

$$\Longrightarrow \Delta D_{disl} = \Delta D_{migr} = 2.25\,\text{nm}$$

$$\frac{\Delta D_{disl}}{D_0} = 7.5\% \qquad \text{(contribution of disl. plasticity)}$$

$$\beta\frac{\Delta D_{migr}}{D_0} = 3\% \qquad \text{(contribution of GB migration)}$$

$$\Longrightarrow \frac{7.5\%}{\varepsilon_{plastic}} \Longrightarrow \frac{7.5\%}{19\%} \approx 40\% \qquad \text{(rel. share of disl. plasticity)}$$

$$\Longrightarrow \frac{3\%}{\varepsilon_{plastic}} \Longrightarrow \frac{3\%}{19\%} \approx 15\% \qquad \text{(rel. share of GB migration)}$$

$$100\% - 40\% - 15\% = 45\% \qquad \text{(rel. share of GB-mediated plasticity)}$$

Intriguingly, the large fraction of GB area per grain volume, inherent to NC materials, contributes significantly to overall strain.

Finally, it is referred to the microstructural issue of the slightly elongated grains and the possible influence on the mechanical response of varying orientation of the elongated grains with respect to the loading direction. This was examined with additional compression tests, shown in Fig. 4.6. Regardless of the initial intensity distribution, all samples show the tendency to redistribute the INT in the same way, forming a similar six-fold symmetry. By applying the SLM to analyze the samples with growth direction oriented perpendicular to the X-ray beam, unexpectedly a rather equiaxed grain shape is deduced (h)-(i). A closer look at the orientation map in Fig. 4.1(c) reveals distinct orientation changes along the columnar grains indicating a considerable substructure of these grains. In fact, the substructure defines the coherent scattering domain size probed by XRD. Therefore, is is concluded that for the materials response the orientation of the columnar grains is less relevant but the random substructure is the determinant microstructural parameter.

Some of the findings described above have already been mentioned or observed in the literature. Nevertheless, there is still an ongoing debate whether dislocation plasticity and/or GB-mediated deformation mechanisms are governing the deformation behavior of NC metals. Indeed, texture evolution, dislocation plasticity and grain growth at large plastic strains have been found in NC NiFe alloys [Fan et al., 2006a,b]. On the other hand, GB-mediated plasticity leading to grain rotation and growth was observed in NC Ni [Shan et al., 2004; Wang et al., 2008]. It was suggested that the rotation of individual grains may lead to multigrain agglomerates by formation of small angle GBs and incomplete grain coalescence which is interpreted as grain growth. Last but not least, stress-driven GB migration in NC Al films has been observed [Rupert et al., 2009; Legros et al., 2008] and discussed as a deformation mechanism accommodating considerable plastic strain without the need of dislocation motion. However, so far it was not possible to identify and separate the contributions of the relevant mechanisms as well as allocate them to specific stress and strain regimes.

This study successfully demonstrates that the deformation behavior of NC Ni entails a characteristic sequence of different deformation mechanisms (Fig. 4.22): A crossover from elastic and GB-mediated accommodation processes to coexistence of GB shear and slip as well as dislocation glide and stress driven grain growth at large strains. Based on the excellent resolution and statistics of the experiment, the sequence as well as the individual shares of deformation modes could be discriminated with unprecedented clearness. To what extent this behavior can be assigned to NC metals in general will be discussed in the following sections and in the comprehensive discussion (see chapter 6).

4.3.2. Influence of Loading Condition

In the following, the mechanical behavior of the NC Ni SCS is discussed and compared to the COMP sample. The shear-dominated deformation of the SCS geometry and the downscaling of sample size was already assessed by FEM in Ref. [Ames et al., 2010] for different materials. The downsizing is a prerequisite for testing NC samples which usually cannot be fabricated at large scales. In order to probe the shear-dominated materials response emerging in the narrow sample slit by *in situ* XRD, an appropriate setup is required. Solely with the unique characteristics of beamline ID15A at the ESRF, combining high energy and microfocus of the X-ray beam, it is possible to probe the materials response during loading solely within the slit of the sample, where shear deformation is dominant.

Based on examination of the optical images of the camera and their analysis by DIC, it is evidenced that the macroscopic deformation of the SCS is localized to the slit, as a result of the reduced thickness ($\approx 100\,\mu m$) compared to the bulk ($\approx 550\,\mu m$). In fact, the macroscopic deformation measured by the relative displacement of the upper to the lower sample half via DIC revealed dominant shear deformation as lateral and longitudinal displacement during loading are of equal magnitude (cf. Fig. 3.3(b)).

The effect of the external loading condition on the deformation state measured by XRD was shown in Figs. 4.8 and 4.11(b) indicating a constant in-plane rotation of the main strain axes of about $5.5°$ in counter-clockwise direction. It is noted that the maximal compressive lattice strain of the SCS is increased by approximately 11% compared to the compression experiment, which may result from lower grain size of the SCS ($D =$

20 nm) relative to the COMP ($D = 30$ nm). Moreover, severely increased maximal tensile lattice strains are observed (see Table 4.1). Comparing SCS and COMP, tensile-over-compressive-strain ratios of 0.40 and 0.30 are estimated, respectively. A simple approach similar to the maximum shear stress theory from Tresca ($\varepsilon_\tau = (\varepsilon_t - \varepsilon_c)/2$, [Gross et al., 2005]) is used to calculate the shear components from the main strain components, yielding shear components for the SCS of $\varepsilon_{111,\tau}^{SCS} = 0.62\%$ and for the COMP $\varepsilon_{111,\tau}^{COMP} = 0.51\%$ for the (111) lattice strains at the maximum lattice strain levels (transition from regime (II) to (III): $\varepsilon \approx 6\%$). Since for the SCS the lateral tensile strain remains constant and the compressive strain decreases in regime (III), whereas both strain values decrease for the COMP, the difference of the shear component is even more pronounced for higher strains: $\varepsilon_{111,\tau}^{SCS} = 0.60\%$ and $\varepsilon_{111,\tau}^{COMP} = 0.46\%$ ($\varepsilon = 20\%$, see also Table 4.1).

The higher shear component in the SCS is regarded to be the reason for the occurrence of the sharper texture (see also Fig. 4.9(d) and the calculated texture ratio of 1.58 for the SCS and 1.44 for the COMP). The observed intensity redistribution corresponds to a typical compression texture for coarse-grained FCC metals [Gambin, 2001], where conventional dislocation plasticity is dominant. The enhanced shear component in the SCS obviously gives rise to a sharper deformation texture, suggesting that dislocations can be easier (nucleated and) activated in the SCS compared to the COMPS. A recent study on NC Pd-10Au SCS [Ames et al., 2012] underlines the ability of shear-dominated deformation to allow for accommodation of large plastic strains, while similar Pd-10Au samples under compressive loading lack in plastic strain accommodation and fail by crack formation and propagation (cf. section 4.2.3).

The specific plasticity of the SCS entails ϕ-dependent changes of other peak parameters: At the same time when texture strongly emerges, the IBR decreases particularly at ϕ angles of maximum INT (compare Fig. 4.9 and Fig. 4.12). Especially the (111) planes are affected by the intragranular plasticity, since the (111) planes are the primary slip planes in FCC metals. Particularly, in the ϕ directions, where the (111) INT is maximal, the SLM analysis, based on the separation of Lorentzian and Gaussian contributions to overall peak shape, fails and yields unreasonable (hkl)-dependent grain shapes, as shown in Fig. 4.14(a). Comparing the (111) INT (Fig. 4.9(a) and (d)) and IBR distributions (Fig. 4.12(a) and (d)) at maximal deformation for SCS and COMP, it is obvious that the shear-controlled dislocation plasticity, observed for the SCS, has a dominating effect on the IBR. This is indicated by the impression of the six-fold symmetry to the IBR, which is not observed during compressive loading, where consequently, no (hkl)-dependent grain shape evolution is obtained. This shear-affected impact on the IBR seems to be more dominant than peak shape changes due to isotropic grain growth or grain shape changes following from dislocation plasticity. To close the line broadening analysis, it is reverted to the pristine IBR. It is reasoned that the strong decrease of IBR in the directions of preferential dislocation glide, and concomitant maxima of INT (Fig. 4.13), suggest, that no dislocations pile-up and leave any debris in the grain interior. Rather, they are absorbed by the surrounding GB network [Derlet et al., 2003a].

As grain size analysis from XRD line broadening failed, the grain size from ACOM/TEM has to be considered (Fig. 4.15), being aware of the reduced statistics. The ellipticity of the in-plane grain shape, relating the major to the minor axis of the ellipse, measures 1.30 for the deformed SCS, and thus is more pronounced than for the COMP (1.15). Thereby, the enhanced role of dislocation plasticity is independently verified by grain shape evolution in addition to the stronger texture formation.

Furthermore, the increased amount of twins in the deformed SCS microstructure (Fig. 4.15(a)), which was neither observed for the deformed COMP nor for the undeformed counterparts, indicates an increased role of partial dislocations in NC Ni relative to full dislocations, since partial dislocations are suspected to be responsible for the formation of deformation twins [Zhu et al., 2012]. The mechanical behavior of NC Ni dominated by extended partials was also predicted by MD simulations [Van Swygenhoven et al., 2004]. One can speculate why the twins are observed in the deformed SCS microstructure but not in the COMP: It is obvious that the altered loading condition and the well-defined deformation geometry in the SCS could account for the twins. On the other hand, twin formation was also found to be very sensitive to strain rate [Roesner et al., 2004], which was of the same order for SCS and COMP. However, strain rate locally increases significantly when crack formation emerges in the slit. Therefore, another plausible argument is that twin formation only occurs simultaneously with or exactly before crack growth in the region of crack propagation along the slit. Please note again, that the TEM lamella was directly extracted at the fracture site. Additional ACOM/TEM analysis, further away from the fracture site, could shed light on this aspect. Unfortunately, the evolution of twins (in terms of a twin fault probability [Warren, 1959; Klug, 1974]) could not be revealed by XRD via relative peak shifts due to the strong elastic grain interactions among different (hkl) families during loading.

To conclude, compared to COMP, several findings imply an increased relative contribution of dislocation-mediated plasticity to overall deformation for the SCS: (i) Higher shear component, (ii) stronger texture formation, and (iii) a more elliptic grain shape after deformation. For the COMP the relative contribution was estimated to be about 40%, while for the SCS, a similar approach that uses the grain shape change measured by ACOM/TEM, yields a higher value around 50%. This insight is even more intriguing, when the difference in grain size is recalled (SCS: $D = 20\,$nm, COMP: $D = 30\,$nm), from which rather a decrease of intragranular dislocation plasticity would be expected for the SCS.

4.3.3. Alloying Effects in PdAu and Differences to Pure Ni

The aim of this section is to identify deformation mechanisms for PdAu alloys ($D \approx 10$ - $15\,$nm) from *in situ* compression experiments and to investigate composition-dependent trends, particularly in which way the contributions of individual mechanisms are modified by alloying. Moreover, the behavior is compared to the behavior of Ni ($D \approx 30\,$nm), which was discussed in detail in section 4.3.1.

First of all, the macroscopic compressive stress-strain curves are compared and discussed. The PdAu alloys show a pronounced strain hardening behavior (Fig. 4.16) instead of a macroplastic plateau as seen for Ni (Fig. 4.2(a)). For example: Pd-30Au with a yield strength $\sigma_y < 1\,$GPa, almost reaches $2\,$GPa after $\varepsilon_c = 28\%$. Similar evolutions are found for Pd-50Au, and also for Pd-70Au however with a significantly reduced overall stress level. The curve of Pd-10Au deviates with the onset of crack formation ($\varepsilon_c > 7\%$). Similar trends, as found for the macroscopic stress over compressive strain, are also observed for the evolution of elastic lattice strain (cf. Figs. 4.16 and 4.17(a)): The lattice strain continuously increases, surprisingly also in the macroplastic regime (except sample Pd-10Au). Therefore, it is argued that the elastic lattice deformation cannot be easily released by plastic events, indicated by the ever ongoing increase instead of saturation. The geometrical constraint to dislocation plasticity in form of small grain sizes of the PdAu alloys ($D \approx 10$ - $15\,$nm) could be one apparent reason. On the other hand, solutes could segregate to the GBs impeding interfacial deformation modes, such as GB migration, GB shear, or grain rotation. Obviously, the prevention of plastic deformation is most severe in Pd-10Au. In this case, the stored deformation energy can only be released by crack formation and propagation.

No pronounced alloying effect on the absolute levels of lattice strains is observed. Certainly, the qualitative behavior of Pd-10Au deviates from the other alloys starting at the onset of crack propagation, as also observed for the macroscopic stress-strain curve. Comparing the lattice strains and macroscopic stresses it is clearly noticeable, that the Pd-70Au exhibits a distinctly lower stress level compared to Pd-30Au and Pd-50Au, although lattice strain is almost as high as for the other alloys.

To build bridges between lattice strain and macroscopic stress, the Young's modulus and its alloy dependence are considered. Certainly, such compression experiments are inappropriate to measure exact Young's moduli from stress-strain curves. Thus, a different approach is pursued: The Young's modulus is calculated by dividing the macroscopic stress by the (111) lattice strain evaluated in loading direction ($\phi = 90°$), see Fig. 4.23(a). Averaging the calculated Young's moduli, in the arbitrary range between 5% and 20% strain, yields similar values for the Pd-10Au, Pd-30Au, and Pd-50Au alloys ($E \approx 128\,$GPa), and a significantly reduced value for Pd-70Au ($E = 101\,$GPa). Accessing literature [Beck, 1995] reveals approximately a plateau for the Young's modulus of bulk PdAu alloys ($E \approx 120\,$GPa) or even slightly increasing values for Au contents $c_{Au} <\approx 60$ at%. For higher Au contents a severe decrease towards pure Au ($E \approx 79\,$GPa) was found (Fig. 4.23(b)). Qualitatively the calculated values agree with the trend from literature, since in both cases a strong reduction of E for Au contents above 60 at% is observed. This decrease of E can account for the observed reduction in strength for the Pd-70Au sample, which in fact has an Au content as high as 77 at% revealed by EDX (cf. Table 3.2). Of course, the calculated Young's moduli are (hkl)-dependent, and the chosen (111) values yield the largest moduli (least lattice strain, cf. Fig. 4.2(b)). However, the selection of (hkl) should only influence the absolute values, but not alloy-dependent changes. Furthermore, errors resulting from the offset angle between plane normal and loading direction (that is θ) are the lowest for (111). The qualitative consistence of calculated values and these from literature, helps to explain the experimental observations.

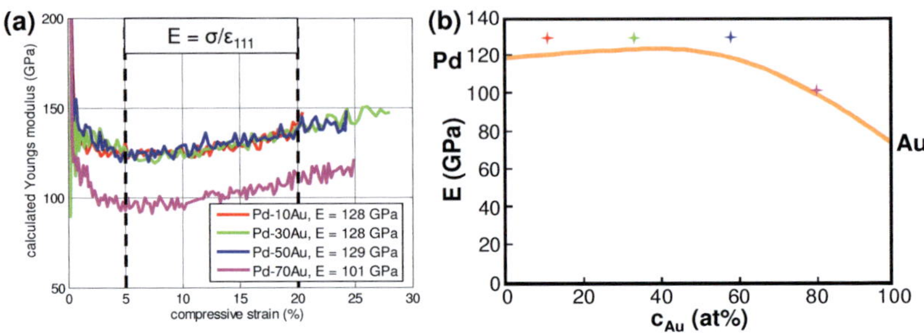

Figure 4.23.: (a) First principle approach to calculate Young's modulus by dividing macroscopic stress by (111) lattice strain. The experimental values, averaged between 5% and 20% compressive strain are listed in the legend and (b) superimposed to Young's modulus data of polycrystalline PdAu alloys adapted from Ref. [Beck, 1995]. The exact Au contents of the compression samples were measured by EDX (see section 3.5.2).

The PdAu alloys show a pronounced evolution of peak asymmetry during loading (Fig. 4.17(c) and (d)), which can be explained as follows: Owing to the high lattice strains ($\varepsilon_{hkl} > 1\%$), pronounced differences among various (hkl) strains develop during deformation, based on different elastic compliances. As a result, elastic grain interactions arise, involving accommodation processes in GBs and triple lines. Limited accommodation processes (e.g. due to steric hindrance to GBS [Hahn et al., 1997]) yield backstresses on the grain interior, causing a premature arrest of the shifting peak of the corresponding, most strained, diffracting lattice planes and thereby, an asymmetric peak shape emerges. For the detailed discussion on the origin of the deformation-induced peak asymmetry and its reversal based on the activation of dislocation-mediated plasticity, the reader is referred to section 5.3.1.

Similar to lattice strain, it is remarkable that the asymmetry values do not saturate in the macroplastic regime, but further increase with increasing strains, which clearly differs from the dislocation-controlled reversal of peak asymmetry observed for Pd thin films (see chapter 5, e.g. Fig. 5.13 or Fig. 5.15). Thus, the non-reversing asymmetry of the IGC PdAu alloys under compression underlines, in addition to the non-saturating lattice strain, how difficult it is to activate intragranular mechanisms in 10 nm-sized grains. On the other hand, considering the ϕ-dependent asymmetry evolution, the trends of $A > 0$ for a compressively strained lattice and $A < 0$ for a tensile strained lattice are consistent for bulk samples and thin films.

Clear alloying effects are observed for the deformation-induced texture evolution (Figs. 4.18 and 4.19 and Table 4.2) and grain growth (Fig. 4.21, Table 4.3). With increasing Au content in the PdAu alloys, the introduced texture parameters, relating the ϕ-dependent INT peak maxima and minima of a particular (hkl) component, increase. The observed texture is similar to typical textures known from CG FCC metals under the same loading conditions [Gambin, 2001], where classical dislocation-based mechanisms are dominant. Therefore, one may argue that dislocation activity becomes more relevant in alloys with higher Au contents.

Recently, it was shown theoretically by a formalism using concentration-dependent embedded-atom method potentials of pure Pd and Au, that both the intrinsic stacking fault energy and twinning fault energy notably decrease with increasing Au content [Schaefer et al., 2011]. In fact, the intrinsic stacking fault energy γ_{sf} decreased by a factor > 4 from pure Pd to pure Au. As a result, for $D = 15$ nm the simulations show stronger increases of partial dislocation and stacking fault densities during loading for higher Au contents. It was also shown that both, the stable and unstable stacking fault energy must be consulted for the discussion of dislocation-based plasticity in NC metals [Van Swygenhoven et al., 2004]. However, since the unstable stacking fault energy, γ_{usf}, is fairly constant over the complete alloy system, the further discussion is based on the stronger varying intrinsic stacking fault energy. In section 2.2.1, it was discussed that below a critical grain size partial dislocations can be nucleated more easily compared to full dislocations [Chen et al., 2003]. A leading partial is emitted from a stress concentration within the GB network. Dependent on the stacking fault energy, a second, trailing partial can follow, yielding a full dislocation, or the leading partial is absorbed at the opposite GB before a trailing partial can emerge [Yamakov et al., 2001; Derlet et al., 2003b]. One can argue, that the alloy-dependent reduction of γ_{sf} allows for a wider splitting of leading and trailing partial dislocations, and hence, it is more difficult in Au rich alloys for dislocations to cross-slip, as only full dislocations can cross-slip. Consequently, the dislocations remain on their primary slip plane. Since glide occurs mainly in preferred orientations with high Schmid factors, a stronger deformation texture emerges for alloys with higher Au contents.

On the other hand, in Pd rich alloys, it should be easier for dislocations to cross-slip, since splitting of leading and trailing partial is narrowed. Cross-slip was shown to be an essential mechanism for the absorption of dislocations in GBs [Bitzek et al., 2008]. However, cross-slip can only occur at GBs in hydrostatic tensile state. If however, areas of solute depletions and enrichments alternate along a single GB, as it was shown in Fig. 3 of Ref. [Schaefer et al., 2011], the areas of hydrostatic tensile and compressive stresses alternate likewise, and probably impede cross-slip and absorption of dislocations. Indeed, the simulations show a stronger Au depletion in the GBs for alloys with low Au contents. Since the atomic radius of Au is smaller than that of Pd ($r_{Au} = 135\,\mathrm{pm}$ and $r_{Pd} = 140\,\mathrm{pm}$ [Slater, 1964]), generally a stronger Au depletion should lead to more compressive stress states in the GBs. Thereby the higher γ_{sf} of Pd rich samples would enable cross-slip, whereas compressive stress states in GBs would retard the absorption of dislocations.

Furthermore, according to Ref. [Gambin, 2001], the ratio of (h00) to (hhh) texture components are influenced by the stacking fault energy with the trend of stronger (h00) components for lower stacking fault energies (in this case, higher Au contents). In CG metals, the effect is again attributed to the ability of dislocations to cross-slip [Brown, 1961]. The herein tested alloys show the same trend of increasing ratio of averaged (200) to (111) peak intensities for increasing Au contents, respectively decreasing γ_{sf}, displayed in the last column in Table 4.2. Consequently it is argued, that also in NC metals a high γ_{sf} can lead to enhanced cross-slip and therefore more pronounced (hhh) texture components compared to (h00). Please note that apart from this trend, all texture components generally increase with increasing Au content, manifesting enhanced overall texture evolution.

To conclude, two points are denoted: (i) Dislocation plasticity is generally more pronounced in alloys with high Au contents (low γ_{sf}), which was inferred from stronger (hkl)-independent texture evolution, and (ii) the composition-dependent trends for dislocations to cross-slip and to be absorbed at GBs may balance the opposite trend to be repelled from the GBs by local compressive stress states. Finally it must be noted, that the INT distributions obtained from the Debye-Scherrer rings only yield in-plane information, and therefore cannot be directly compared to the full texture analyses in Refs. [Wassermann, 1962; Barrett and Massalski, 1966; Gambin, 2001]. However, although some information is lacking, the *in situ* XRD data can be reconciled with literature, yielding similar, qualitative trends.

In the next step, grain size and shape as well as their evolution are discussed. Initially, all alloys exhibit a perfectly equiaxed in-plane grain shape, see Fig. 4.21. The relative increase in grain size observed after $\varepsilon_c =$ 20% is independent of the initial absolute grain size, but alloy-dependent with the strongest relative grain size increase for highest Au contents (see Table 4.3). The fairly preserved equiaxed grain shape, again underlines the diminishing role of dislocation plasticity for the very small grain sizes, since prevailing dislocation activity would yield an elliptic grain shape (non-uniform flow [Ashby and Verrall, 1973]). Instead, stress-induced GB migration might play a more significant role, whereupon the initial grain shape is preserved, while grains grow isotropically, as shown in Ref. [Rupert et al., 2009].

In Ref. [Lohmiller et al., 2012a], the ratio of background intensity (diffuse scattering) to integral peak intensity was additionally used to analyze the deformation behavior in NC Pd thin films. In the original papers introducing this parameter and analyzing its interplay with microstructural changes [Ungar et al., 2005, 2007], the increased ratio was interpreted as a result of increasing vacancy concentration during plastic deformation. Transferring this parameter to NC microstructures, one can argue that the increasing ratio indicates an increased ability of GB accommodation, based on rearrangement of excess-volume (e.g. by atomic shuffling [Van Swygenhoven and Derlet, 2001] or similar point defect-related diffusive processes) in the enlarged GB network. Coming back to the tested PdAu alloys, during loading this ratio increases most pronouncedly for highest Au contents, as shown in Fig. 4.24. The ratio was calculated from the sum of the (111), (200), and (220) INTs and the underlying background intensities averaged over the full Debye-Scherrer rings, in order to be independent of texture effects. Similar to higher test temperature in the Pd thin films [Lohmiller et al., 2012a], the higher Au content could account for the improved ability for individual rearrangement processes and thereby, could also explain the enhanced GB mobility for alloys with higher Au contents, correlating with the most pronounced relative grain size increase. The least increase of the ratio, observed for Pd-10Au, correlating with least relative grain size increase, can be again attributed to crack formation and the inhibited accommodation of plastic deformation. It is pointed out, that the ratios are normalized to their initial values. The approach does not allow to obtain information from the absolute values and compare them among the different alloy compositions, but only the relative changes during deformation can be followed[1].

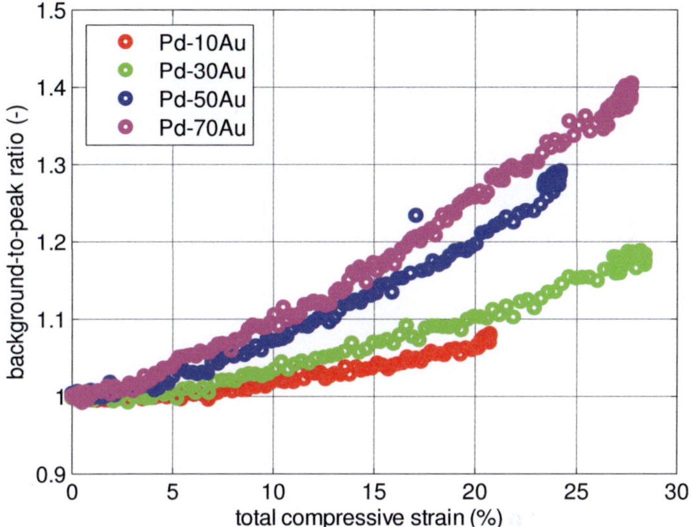

Figure 4.24.: Ratio of background to peak intensity averaged over (111), (200), and (220) INTs and their underlying background intensity. Additional averaging over the complete Debye-Scherrer rings should eliminate effects due to texture formation. Stronger increases are observed for higher Au contents.

The pinning of GBs by solute atoms could also influence their mobility. However, neither a dependence on solute concentration, nor a dependence on initial grain size is observed. MD simulations [Schaefer et al., 2011] have shown that the concentration of solutes in GBs can distinctly differ from the bulk concentration, and furthermore in a single GB a solute depletion can occur adjacent to a solute enrichment. Therefore, it is challenging to identify an alloying effect on GB pinning from which alloy-dependent relative grain growth could be derived. It is

---

[1]The definition of the 2θ range used for the calculation of the background intensity is arbitrary. Basically, for all alloys the same range can be defined or the 2θ range could be scaled with the individual width of the corresponding peak of each PdAu alloy.

speculated, that rather an enhanced vacancy formation and rearrangement of excess-volume may account for the observed alloy-dependence.

Finally, the compressive behavior of the PdAu alloys is compared to Ni. Seizing the grain size-dependent behavior, discussed by means of Figs 2.1 in section 2.2.6, it is pointed to the difference in absolute grain sizes: PdAu alloys ($D \approx 10$ - $15\,$nm) are rather at the lower bound of the transitional regime, whereas Ni ($D \approx 30\,$nm) is located in the middle of the transitional regime.

The yield limits of the PdAu alloys are distinctly lower ($\sigma_y < 1\,$GPa, see Fig. 4.16) than the yield limit for pure Ni ($\sigma_y \approx 1.5\,$GPa, Fig. 4.2(a)), although or just because grain sizes of the alloys are in the range of 10-15$\,$nm, while Ni has an initial grain size of 30$\,$nm. On the other hand, during macroplastic deformation, the strain hardening behavior is much more pronounced in the PdAu alloys while Ni rather reveals a macroplastic plateau. The same behavior is observed for the lattice strain evolution: For Ni, the lattice strain saturates, and even decreases during macroplastic deformation, while lattice strain continuously increases in the alloys. These trends demonstrate the restriction on stress release inside the small sized grains.

Comparing the correlation of absolute macroscopic stress and lattice strain of PdAu alloys and Ni it is found that, although macroscopic stresses of Ni are higher than of PdAu alloys, the grain interior of the alloys bears more elastic strain (Ni: $\varepsilon_{111,max} = -0.8\%$ and PdAu: $\varepsilon_{111} > -1.2\%$). This again can be explained with the different elastic constants which are significantly higher for Ni ($E \approx 210\,$GPa [Zacharias, 1933]) than for Pd and Au. Nevertheless, the (111) elastic lattice strain values of $> 1.2\%$, which are still increasing for $\varepsilon_c > 20\%$, are remarkably.

Besides the differences in lattice strain evolution, further indications reveal a change to stronger dislocation impediment but enhanced

GB-mediated deformation for PdAu alloys compared to Ni. The texture formation is clearly reduced for the alloys, indicating a reduced activity of dislocations. Furthermore, grain shape remains fairly equiaxed, and in addition, the relative grain size change is more pronounced, pointing to an increased activity of GB migration. This argument is supported by the overall decrease of INT for PdAu alloys and the related discussion of the background-to-peak ratio (cf. Fig. 4.24), whereas for Ni the normalized INT fluctuates around 1.

To conclude this section, the deformation behavior in PdAu alloys with $D \approx 10\,\text{nm}$, clearly differs from pure Ni with $D \approx 30\,\text{nm}$ under the same loading conditions. Dislocation activity is efficiently hindered due to the geometrical constraint of the very small grains. This is demonstrated on the basis of several XRD parameters independently: (i) The generally weaker texture formation in the alloys, (ii) the preserved equiaxed grain shape, (iii) the continuously increasing lattice strain, owing to the increased hindrance of releasing elastically stored energy by plastic processes, and (iv) the non-reversing peak asymmetry. Furthermore, among the different PdAu alloys a composition-dependent behavior is observed, with the trend that alloys with higher Au contents show stronger texture formation and most pronounced relative grain size increase. Plasticity in form of dislocation-mediated processes and GB migration seems to be promoted in alloys with higher Au contents, whereas for lower Au contents, GB-mediated deformation (e.g. GBS) must undertake an increased contribution to overall deformation. However, the ability of GBS (and accompanying diffusive or rotational processes) to accommodate plastic strain is maybe limited at the given strain rate and temperature, explaining the rather brittle materials response for lowest Au contents. To conclude, a maximal brittleness is observed for Pd-10Au and increasing ductility for higher Au contents.

4.4. Summary

The deformation behavior of bulk NC metals and alloys, examined by synchrotron-based *in situ* compression and shear-compression testing and sophisticated peak shape analysis in combination with complementary ACOM/TEM analysis, can be summarized as follows:

- Not only the coexistence of dislocation plasticity, grain growth and interfacial deformation modes could be proven, but it was also possible to assign them to distinctive strain regimes. In addition, the relative contributions to overall deformation were estimated quantitatively for the first time.

- The deformation behavior of NC Nickel ($D \approx 30\,\mathrm{nm}$) can be classified into three regimes: (I) Inhomogeneous elastic lattice straining and GB accommodation, followed by (II) dislocation plasticity, which was inferred from texture evolution, and (III) stress-driven grain growth. The deformation is governed by a succession of different, partly overlapping mechanisms.

- The relative contributions to overall deformation after 20% compressive strain can be estimated to 40% dislocation-mediated plasticity, 15% stress-driven GB migration, and 45% GB-mediated deformation.

- Shear-dominated deformation of NC Ni promotes dislocation-mediated plasticity.

- The mechanical behavior of PED Ni is rather unaffected by the orientation of initially elongated grains with respect to the loading direction, but the random subdomain structure seems to be the determinant microstructural parameter.

- Compared to Ni, dislocation plasticity is considerably restrained for PdAu alloys ($D \approx 10 - 15\,\mathrm{nm}$). This observation was evidenced by several parameters individually: (i) The generally weaker texture formation, (ii) the preserved equiaxed grain shape, (iii) the continuously increasing lattice strain, and (iv) the non-reversing peak asymmetry. Instead, GB-mediated deformation carries an increased contribution to overall deformation.

- For the deformation-induced texture formation and for grain growth, alloy-dependent behavior is observed with more pronounced changes for higher Au contents.

- For alloys with low Au contents, the relative contributions change for the benefit of GB-mediated deformation (e.g. GBS) on the expense of dislocation-mediated deformation and GB migration.

- Overall the investigated alloys reveal a maximal brittleness for Pd-10Au and an increasingly ductile behavior for higher Au contents.

5. The Tensile Deformation Behavior of NC Pd and PdAu Thin Films

5.1. Introduction

In the following chapter, NC Pd and PdAu thin films adherent to compliant polyimide substrate were tested by a synchrotron-based *in situ* XRD tensile testing technique, in order to obtain further insight in the evolution and succession of different deformation mechanisms. Magnetron sputter-deposition (see section 3.5.3) with base pressures close to ultra high vacuum conditions is employed, in order to fabricate very pure and clean samples. Co-sputtering allows to create alloys over the entire binary PdAu alloy system. The principle deformation behavior of NC thin metal films is investigated with pure Pd. Furthermore, alloying effects are investigated by adding Au to Pd, forming continuous miscible binary alloys. The solute atoms are possibly able to pin GBs and hence, should lead to reduced initial grain sizes and to decreasing GB mobility compared to the pure material [Millett et al., 2007; Wang et al., 2007]. As a consequence, it is expected that, (i) the stability of NC alloys is enhanced [Koch et al., 2008], (ii) the contributions of GB-mediated deformation mechanisms are reduced, and (iii) a strengthening effect arises [Scattergood et al., 2008]. However, one has to be careful in separating strengthening effects based on solid solution and grain refinement.

In order to deform the thin films up to several percent of plastic deformation without failure, the films were supported by compliant substrate. It was shown in Refs. [Li et al., 2005; Xiang et al., 2005; Lu et al.,

2007] that using compliant substrate allows for delocalized deformation (in case of good adhesion) and thereby defers film cracking to higher strains. Polyimide Kapton E (Du Pont) with a thickness of 50 μm was used, as higher ductility (absence of cracking) is attained with this substrate due to better substrate surface quality as compared to standard polyimide [Lohmiller et al., 2010]. Further advantages of using compliant substrate are (i) fault tolerance, since a single flaw does not cause catastrophic failure, and (ii) investigation of cyclic deformation. The fairly elastically deforming substrate forces the plastically deforming film into compressive strain states during unloading. Thereby, tensile and compressive loads can be applied to the sample during cyclic loading.

In this study, in addition to state-of-the-art analysis of peak position, special emphasis is placed on the evolution of peak asymmetry, which notably develops in the peak profiles of the tested NC Pd and PdAu thin films. The reversal of peak asymmetry during loading and unloading manifests the strain-dependent transition from the extended microplastic regime to dislocation-based macroplasticity. Furthermore, quantitative analysis of grain size and microstrain is performed by applying line broadening analysis (see section 3.4.2). Most of the synchrotron experiments were conducted at the MS beamline of the Swiss Light Source SLS (Villigen, Switzerland) (subsection 3.2.2). In order to investigate the in-plane evolution of peak parameters, especially the formation of deformation textures, individual experiments were carried out at the MPI-MF beamline of the Angströmquelle Karlsruhe ANKA (Karlsruhe, Germany), where the setup is equipped with an area detector (subsection 3.2.3). Texture analysis can help to differentiate between dislocation- and GB-mediated deformation [Ma, 2004], and thereby, untangle different coexisting deformation mechanisms. More details on the experimental setups and data analysis are given in sections 3.2 and 3.3, respectively, and in Refs. [Lohmiller et al., 2012a, 2013].

5.2. Results

The results section is presented in two parts. At first, the results from Pd thin film testing are shown: Cyclic experiments were conducted with increasing maximum strain ε_{max} for each cycle, as well as continuous tests up to high plastic strains $> 10\%$. In the second part, alloying effects on the microplastic material behavior are investigated over a large spectrum of alloy compositions ranging from pure Pd to Pd-72Au with 72 at% Au. Additionally, similar to Pd, a cyclic experiment ($\varepsilon_{max} = 6\%$) is conducted on a Pd-12Au thin film. The *in situ* synchrotron experiments are complemented by ACOM/TEM investigations (see Appendix A for details on the method). Furthermore, SEM investigations of the deformed samples shed light on the deformation morphology of the tested thin films.

5.2.1. Pd Tensile Testing

5.2.1.1. Strain Increase Test

A Pd thin film sample was loaded and unloaded five times, each time with increasing maximum strain, up to $\varepsilon_{max} = 6\%$ after the 5th cycle. The absence of film cracking in the tested sample was confirmed by SEM investigations after tensile testing. The evolution of (111) XRD peak parameters in loading direction ($\phi = 90°$) is shown as a function of true strain ε in Fig. 5.1. The peak position (Fig. 5.1(a)) decreases during loading, as lattice spacing increases responding to the applied tensile load. The initial shift is fairly linear for $\varepsilon < 0.3\%$ and deviates from linearity for larger strains. After unloading of cycle 1, larger 2θ values are reached, since the compliant substrate forces the microplastically deformed metal film into a compressive strain state during unloading, as seen in Fig. 5.1(b). The observed trends are naturally more pronounced in cycle 2. In cycle 3, the 2θ value reaches a plateau while in cycle 4 and 5 the minimum 2θ values are increasing again during further straining. Likewise, the maximum lattice strain is reached in cycle 3 and decreases with further straining. During unloading, the substrate forces the film into a more and more compressive state with increasing ε_{max}. Remarkably, while unloading, the lattice strain evolution strongly deviates from linear behavior, indicative for some share of compressive plastic deformation. As a consequence of irreversible shares of deformation, a hysteresis arises in the lattice strain evolution of each test cycle. The gain of hysteresis width in subsequent cycles suggests an increase of the accumulated plastic deformation (tensile and compressive).

As seen in Fig. 5.1(c), the integral peak breadth (IBR) increases during loading for $\varepsilon > 0.3\%$ and is almost fully reversible after unloading of cycle 1. However, after cycle 2 and 3 irreversible shares are measured. The maximum IBR is reached during the 4th load cycle. Intriguingly, during unloading of cycle 4 and 5, the IBR first decreases as expected, but then increases again during unloading. This is another indicator for plastic compressive deformation during unloading.

The peak asymmetry evolves linearly towards a right-skewed shape during loading in cycle 1 and fully reverses during unloading. During loading cycle 2, a slight deviation from linear behavior is observed. Surprisingly, during cycle 3 the asymmetry reverses during loading and the peak becomes again more symmetric with further straining. This trend is continued during straining in cycle 4 and 5. During unloading, starting from cycle 3, the shape not only fully recovers to symmetric shape, but even becomes left-skewed. For cycle 4 and 5, the left-skewness also shows a reversal similar to the reversal of the right-skewness during loading. This is observed at the same point of unloading, where compressive yielding was deduced from increasing IBR and increasing deviation of lattice strain from linearity. Please note that all other reflections show qualitatively the same overall behavior (see Appendix C).

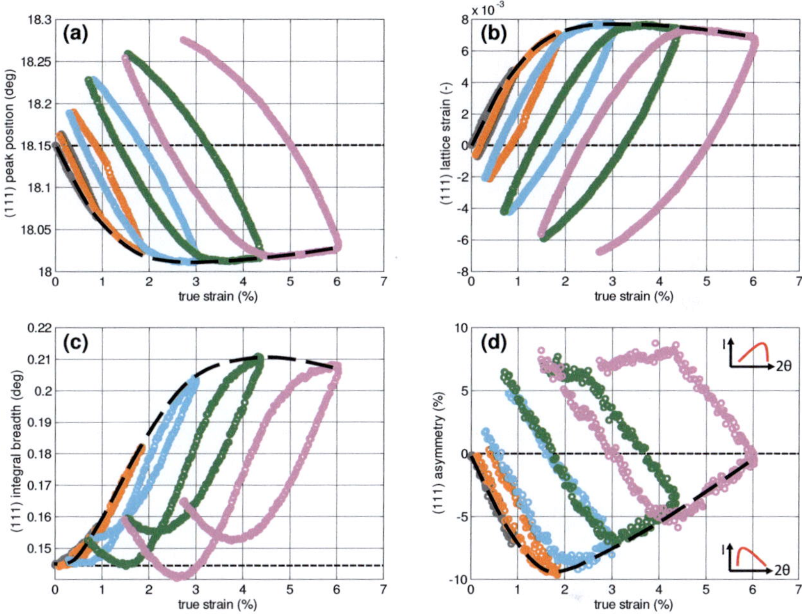

Figure 5.1.: Evolution of (111) XRD peak parameters over ε for a NC Pd thin film. (a) peak position, (b) calculated lattice strain, (c) integral breadth, and (d) asymmetry. Based on their evolution during several load-unload cycles with increasing ε_{max} different deformation modes can be deduced. Note that the use of compliant substrate allows for compressive strain states during unloading. The dashed lines serve as guide to the eye to follow the continuous tensile behavior.

X-ray line broadening analysis is used to separate grain size D and microstrain $<\varepsilon>$. For this purpose the Single Line Method [de Keijser et al., 1982] is used, focusing on a single reflection in order to avoid artifacts of multiple-peak approaches resulting from differently loaded (hkl) planes and pronounced elastic grain interactions in the NC aggregate. Note that in this *in situ* setup with transmission geometry, the diffracting planes of different (hkl) planes experience different load, since the angle between loading axis and scattering vector increases with increasing $2\theta_{hkl}$ (see section 3.4 and Fig. 3.1(b)). Grain size D and microstrain $<\varepsilon>$, evaluated in the unloaded state and averaged over the first five (hkl) families, are shown in Fig. 5.2 as a function of ε_{max} of the previous cycle. Intriguingly, already after the first cycle ($\varepsilon_{max} = 0.9\%$), grain growth and additional microstrain are observed. This trend proceeds with further loading cycles. Consistently, in NC Al films grain growth has been observed after $\varepsilon < 2\%$ [Gianola et al., 2006]. The evolution of grain size and microstrain is not dependent on the selected (hkl) reflections, but is similar for the individual (hkl) reflections, as shown in Fig. C.5 in Appendix C. Therefore, several (hkl) families can be averaged, reflecting the collective relative grain size increase, despite the mentioned issues of multiple-peak approaches, since the SLM decouples these effects by individual analysis of each (hkl) family (cf. the discussion of SLM vs. WH in section 3.4.2 and Fig. 3.10 therein). The averaged, absolute XRD grain sizes and the relative increase in grain size have been confirmed by ACOM/TEM analysis (see Fig. 5.2(a)).

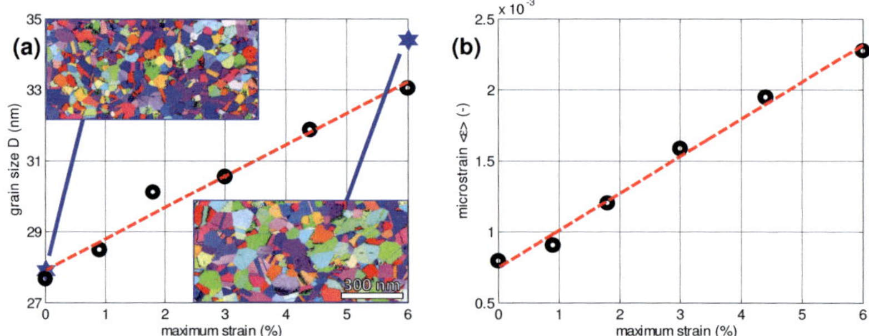

Figure 5.2.: Grain size D and microstrain $<\varepsilon>$ calculated by the SLM [de Keijser et al., 1982] from peak broadening data in the unloaded states. Already after the first cycle ($\varepsilon_{max} = 0.9\%$) grain growth and additional microstrain are measured. This trend proceeds with further cycles. The averaged grain sizes are compared to grain sizes from ACOM/TEM (see insets).

5.2.1.2. Continuous Test

In order to investigate the in-plane evolution of peak parameters ($0° \leq$ $\phi \leq 360°$, not only the loading direction $\phi = 90°$), additional experiments on NC Pd thin films were conducted at the ANKA (subsection 3.2.3), where the setup is equipped with an area detector, similar to the setup used in chapter 4. Instead of 10 reflections in one individual ϕ direction, two complete Debye-Scherrer rings are obtained. With this setup, XRD parameters can be analyzed in any individual ϕ direction, as exemplarily shown in Fig. 5.3 for the calculated lattice strain and the normalized IBR for the (111) reflection. During the continuous tests, lattice strain and IBR in tensile direction ($\phi = 90°$ and $\phi = 270°$) evolve qualitatively in the same way as during the cyclic experiment, shown in Fig. 5.1. When considering the lateral directions ($\phi = 0°$ and $\phi = 180°$), lattice strain is compressive and absolute strain values are distinctly lower than in the tensile direction, as also observed for the IBR.

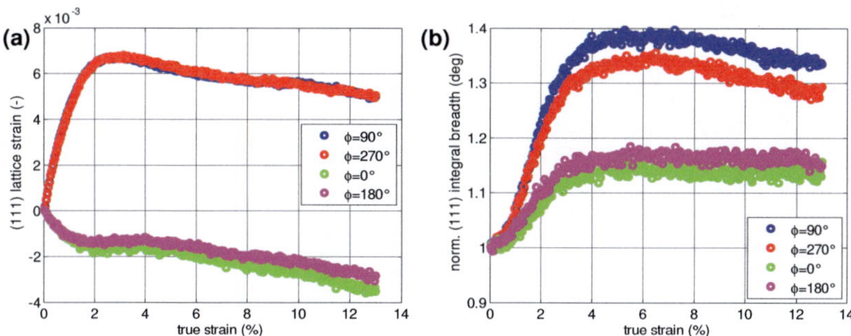

Figure 5.3.: Continuously tested Pd film: (a) (111) lattice strain and (b) (111) IBR over ε for tensile ($\phi = 90°$ and $\phi = 270°$) and compressive directions ($\phi = 0°$ and $\phi = 180°$).

The in-plane evolution of the (111) integral peak intensity (INT) with increasing true strain ε is displayed in Fig. 5.4. Initially, the intensity distribution is elliptic, with intensities almost doubled in vertical direction, compared to horizontal direction (Fig. 5.4(a)). This results from different focusing in horizontal and vertical direction of the provided beam at the MPI-MF beamline, as it is easier to focus and amplify the intensity in vertical than in horizontal direction. Moreover, the ANKA is unfortunately not operated in top-up mode, and hence the provided incoming beam intensity decays over time. However, normalizing the INT to its initial value and furthermore accounting for the decaying incoming intensity over time by introducing a correction term mimicking the time-dependent decay of the incoming beam intensity, the deformation-induced change of the INT can be deduced, as shown in Fig. 5.4(b). For $\varepsilon > 2\%$, the in-plane INT redistributes along ϕ resulting in a weak, yet distinct six-fold symmetry with maxima in tensile direction ($\phi = 90°$ and $\phi = 270°$) and directions thereto rotated by $60°$. INT minima are observed in between.

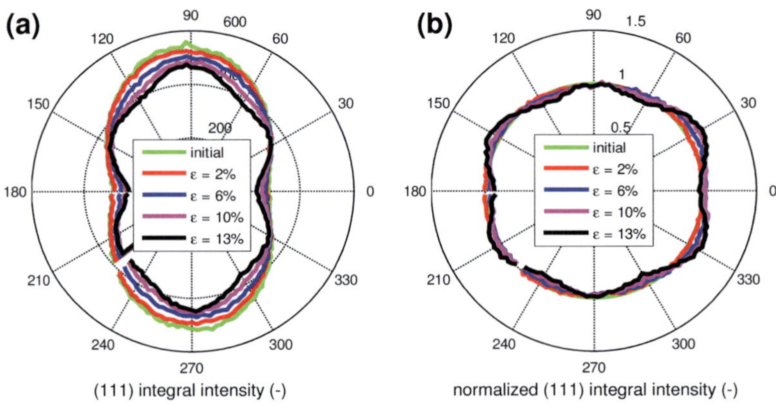

Figure 5.4.: In-plane evolution of the (111) INT. (a) absolute INT and (b) normalized INT to its initial value and corrected for decaying incoming beam intensity.

After tensile testing to $\varepsilon_{max} = 13\%$, the sample was analyzed in the SEM. No cracks were observed in the investigated area, instead a homogeneous shear band pattern spreading the entire sample was found, as shown in Fig. 5.5. The shear bands demonstrate that the NC thin film tends to localized deformation. However, since the film is substrate-supported, localization is restricted and film cracking deferred to higher strains.

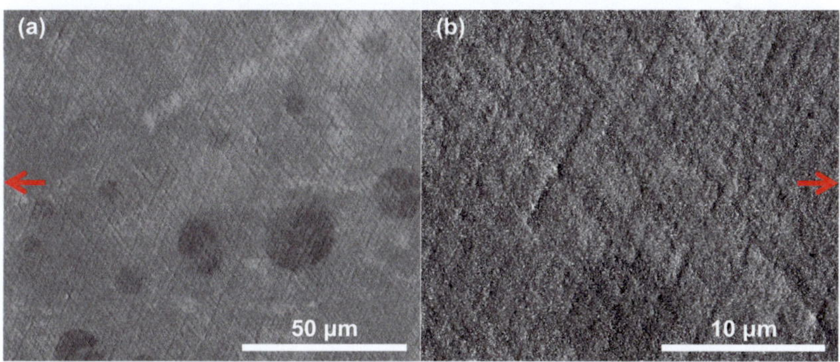

Figure 5.5.: Shear band formation within a NC Pd film after $\varepsilon = 13\%$ observed by SEM. (a) 2 kx (b) 10 kx. The loading direction is horizontal.

5.2.2. PdAu Tensile Testing

Alloying effects on the tensile behavior of NC thin films were investigated by alloying Au to Pd. Details on the fabrication process and the individual sputter parameters are given in section 3.5.3. Due to the continuous miscibility of the PdAu alloy system, no additional peaks appear in the diffraction patterns, but - according to Bragg's Law (Eq. (3.1)) - the peaks of pure Pd shift their 2θ position to lower values, since the lattice constant increases from pure Pd to pure Au ($a_{Pd} = 3.8874\,\text{Å}$, $a_{Au} = 4.0789\,\text{Å}$ [Beck, 1995]). This is shown in Fig. 5.6 exemplarily for the (111) and (200) peaks of investigated alloy compositions.

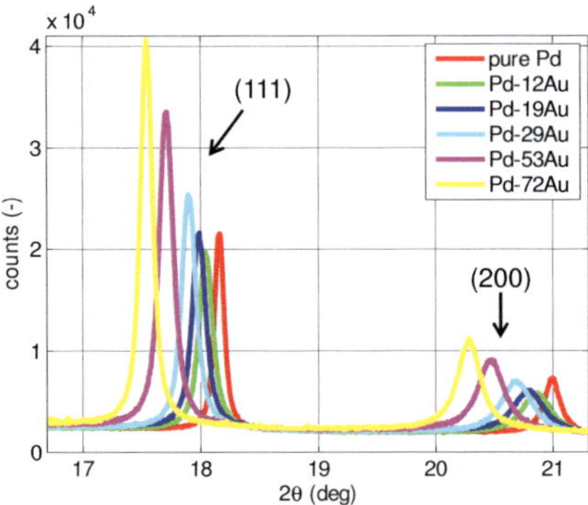

Figure 5.6.: Blow-up of the diffraction patterns recorded for the different alloy compositions in the initial unloaded state. With increasing Au content, the peak positions shift to lower 2θ values.

In Fig. 5.7(a), where the evolution of the fitted $2\theta_{111}$ values of the different alloy compositions are displayed over true strain ε, the different initial values are also visible on the ordinate of the graph, with decreasing 2θ values for increasing Au contents. During tensile loading to $\varepsilon_{max} \approx$ 1.8%, the 2θ positions of all alloys in loading direction decrease, as a result of elastic straining of lattice planes. After unloading, higher 2θ values are reached for all alloys, because, relative to the initial state, a compressive strain state is reached after deformation, as seen in Fig. 5.7(b). Following the evolution of elastic lattice strain over ε, higher lattice strains are measured for alloys with higher Au content. Furthermore, the pure Pd demonstrates the strongest deviation from the ideal elastic slope of 1 (dashed line).

The initial IBR also depends on alloy composition: Alloys with intermediate Au content show largest values, while pure Pd and Pd-72Au exhibit lower values, see Fig. 5.7(c). On the other hand, when the IBRs are normalized to their initial values, these two samples show the strongest increase during straining, and furthermore, clearly irreversible shares after

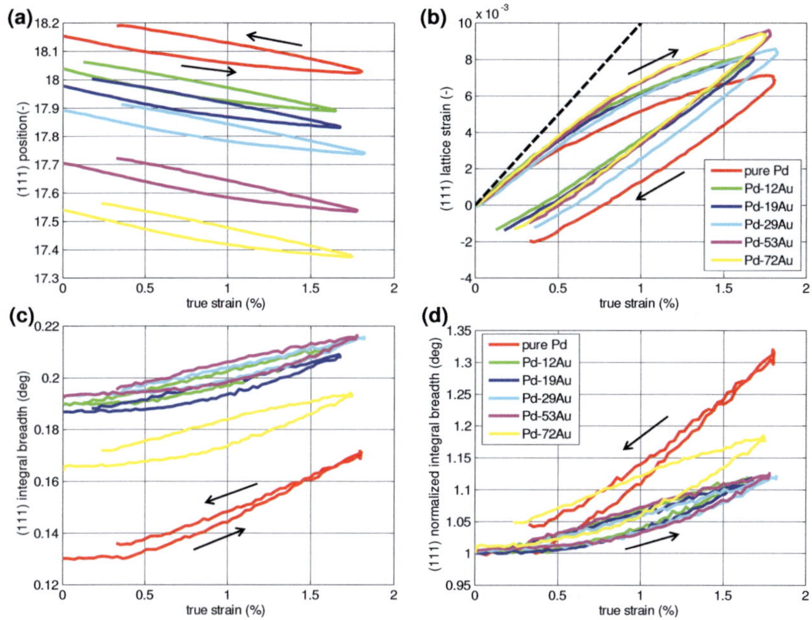

Figure 5.7.: Evolution of (111) peak parameters of PdAu thin films during loading and unloading: (a) peak position, (b) lattice strain, (c) IBR, and (d) normalized IBR as function of true strain ε. The dashed line in (b) represents the ideal elastic slope of 1.

unloading (Fig. 5.7(d)).

For $\varepsilon < 1.8\%$, the evolution of peak asymmetry over ε, shown in Fig. 5.8, does not show a pronounced alloying effect. The alloy behavior is qualitatively similar to pure Pd: The initially fairly symmetric peak develops a pronounced right-skewed asymmetry during tensile loading, which reverses during unloading. However, the absolute change of peak asymmetry at the maximum strain compared to the initial state is most pronounced for pure Pd. The figure legend includes the absolute changes of asymmetry, ΔA, at $\varepsilon = 1.7\%$ compared to the initial values for all samples.

The SLM was applied in order to determine grain size D and microstrain

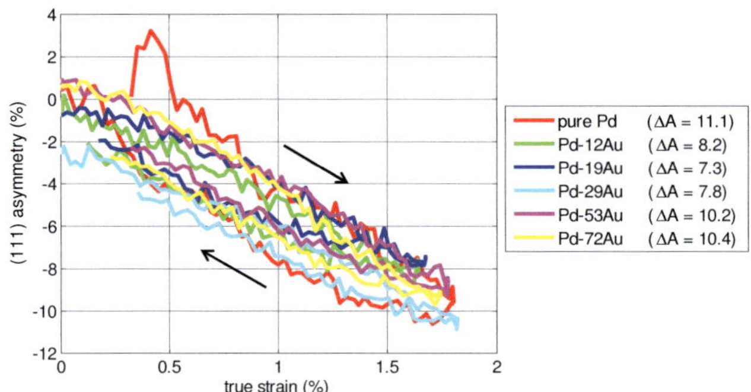

Figure 5.8.: Evolution of (111) peak asymmetry for PdAu alloys during loading and unloading. Smaller absolute changes of asymmetry, ΔA, are observed for alloys with low Au content (see legend for ΔA values.

$<\varepsilon>$ from peak broadening. The individual values of the (111), (200), (220), (311), and (222) peaks are averaged, avoiding (hkl)-dependent effects in absolute grain size. The results for the initial and unloaded state are displayed in Fig. 5.9. In the initial states, the largest D is measured for pure Pd ($D = 30\,\text{nm}$), while all alloys exhibit distinctly smaller D, with smallest $D = 20\,\text{nm}$ for Pd-53Au. On the other hand, smallest $<\varepsilon>$ is measured for the pure Pd film, and all alloys exhibit higher values. A maximum in $<\varepsilon>$ is measured for Pd-29Au, and strongly reduced values for higher Au contents. After ε_{max}, grain growth and additional irreversible microstrain are observed for all samples. The strongest grain size change is observed for pure Pd, and the strongest increase in $<\varepsilon>$ is measured for Pd-53Au, Pd-72Au, as well as for pure Pd. The increase of $<\varepsilon>$ for the intermediate alloys, with already large initial $<\varepsilon>$, is approximately only the half of the increase of the other samples.

For the purpose of investigating alloying effects on the evolution of peak parameters at higher plastic strains, a cyclic strain increase test was conducted on a Pd-12Au film, similar to the test on pure Pd (see section 5.2.1.1 and Fig. 5.1 therein). The evolution of the corresponding (111)

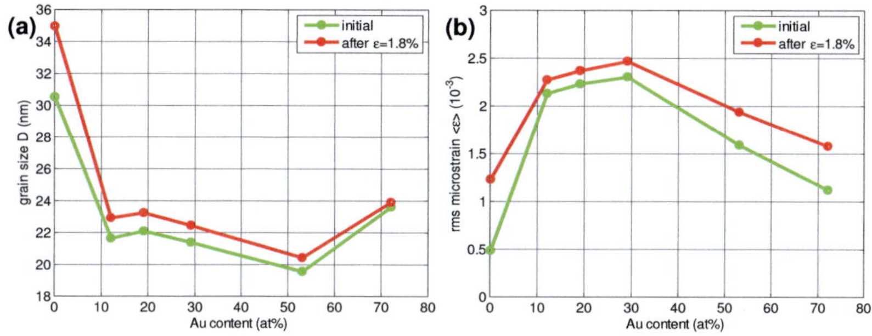

Figure 5.9.: Alloy-dependent results obtained by the SLM: (a) D and (b) $<\varepsilon>$ in the initial states and after $\varepsilon_{max} \approx 1.8\%$.

peak parameters is shown in Fig. 5.10. Generally, the qualitative evolution of peak parameters is similar to pure Pd. However, there a several distinct differences: The peak shift yields an almost 25% higher lattice strain of Pd-12Au compared to pure Pd. Almost 1% (111) elastic lattice strain is measured for the alloy. Furthermore, the initially already larger IBR, increases continuously with increasing ε_{max} from cycle to cycle, while in pure Pd a global maximum is reached in cycle 4 and a lower maximum value is measured in cycle 5. Generally, a lower relative IBR increase during loading is observed for the alloy in each cycle. On the other hand, during unloading the IBR of the alloy does not fall below the initial IBR value as observed for Pd. Both samples exhibit initially symmetric (111) peaks. During loading, qualitatively the same asymmetry behavior is observed with an evolving right-skewed peak and reversing asymmetry with further straining. However, the maximum absolute asymmetry value is slightly lower in the alloy and moreover, the strain at which the asymmetry reverses is clearly larger in the alloy. For $\varepsilon > 5\%$ the asymmetry of the alloyed sample kinks. Simultaneously, a small dip can be observed in peak position / lattice strain.

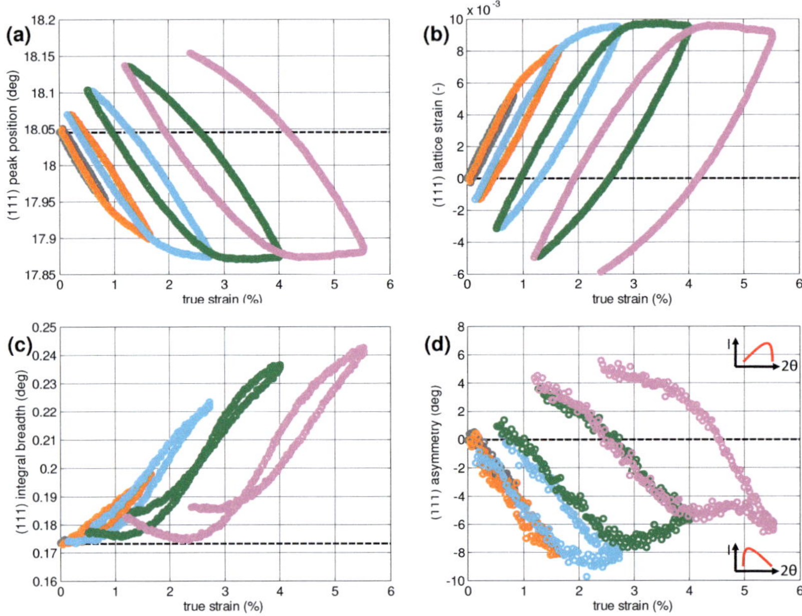

Figure 5.10.: Evolution of (111) XRD peak parameters over ε for a NC Pd-12Au thin film. (a) peak position, (b) calculated lattice strain, (c) integral breadth, and (d) asymmetry. In consequence of the onset of film cracking at $\varepsilon > 5\%$, a dip in position / lattice strain and the rereversing asymmetry are observed.

The small dip in lattice strain and the rereversing asymmetry are consequences of the onset of film cracking. The cracking behavior was investigated *in situ* under an optical microscope during continuous tensile tests, and it was found that film cracking indeed begins at $\varepsilon >\approx 5\%$ for Pd-12Au. The effect of cracking on the peak shape is analyzed in detail in Fig. 5.11. The peak centroid (COM) relaxes twice as much as the peak maximum (POS) from $\varepsilon = 4.9\%$ to $\varepsilon = 5.4\%$, with $\Delta 2\theta_{COM} = 0.008°$ and $\Delta 2\theta_{POS} = 0.004°$. Generally, a film crack in a substrate-supported film causes a local relaxation of the surrounding structure [Xia and Hutchinson,

2000]. This is seen by the overall shifting peak towards the unstrained state. Furthermore, the deformation state of the probed volume (compare beam size of $\approx 500\,\mu m \times 500\,\mu m$ with a crack spacing $> 100\,\mu m$) becomes more inhomogeneous again, indicated by the rereversing asymmetry. It is argued, that the shift of the left flank can only be caused by the grains, representing this flank, however the shift of the right flank can be caused by all diffracting grains, dependent of how strong the lattice strain in each individual grain relaxes. As a consequence, the response to cracking is not a self-similar peak shift, but entails a rereversing asymmetry (increasing A), in addition to reductions in lattice strain.

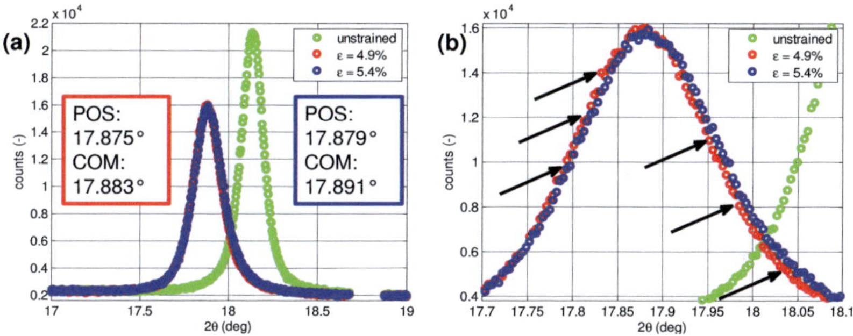

Figure 5.11.: Comparison of peak shape before ($\varepsilon = 4.9\%$) and after ($\varepsilon = 5.4\%$) the onset of cracking. As a result of film cracking, the diffraction peaks shift, according to relaxation of lattice strain. The shift is not self-similar, which can explain the rereversing asymmetry.

The tested Pd-12Au sample was also investigated in the SEM after tensile testing to $\varepsilon_{max} = 5.5\%$ in order to identify the deformation morphology and crack formation. The results are shown in Fig. 5.12. Indeed, a few sporadic cracks are observed with crack distances larger than $100\,\mu m$. Apart from these cracks a fine shear band pattern is present, as it was seen for pure Pd thin films (see Fig. 5.5).

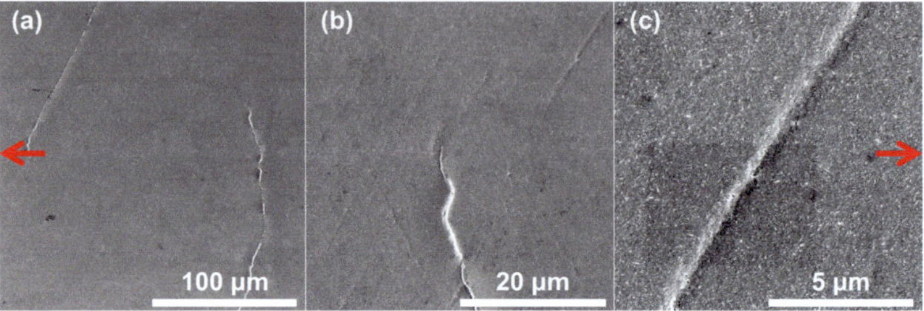

Figure 5.12.: Results from SEM investigations after tensile testing to $\varepsilon = 5.5\%$. The loading direction is horizontal.

5.3. Discussion

The discussion is presented in two parts: (i) The mechanical behavior of NC Pd thin films on compliant substrate investigated by synchrotron-based *in situ* XRD is discussed considering the results from the cyclic strain increase test, as well as from continuous testing following the in-plane evolution of peak parameters. (ii) Alloying effects are discussed by analysis of the composition-dependent behavior of PdAu films and comparison to the behavior of pure Pd.

5.3.1. Succession of Deformation Mechanisms

In order to analyze the interplay of the different XRD parameters in detail and interpret their evolution in the light of microscopic deformation processes, the position, IBR, and asymmetry of the loading data of cycles 1-5 are stitched together in one masterplot (Fig. 5.13) and envelopes are adapted to the data (solid lines). Based on the evolution of the peak parameters two strain-dependent crossovers of deformation modes are identified. The introduction of the following trisection of deformation behavior in NC Pd thin films was proposed [Lohmiller et al., 2012a]: (i) elastic, (ii) microplastic, and (iii) macroplastic. The first regime ($\varepsilon < 0.3\%$) is dominated by elasticity, indicated by the linear change in peak position / lattice strain, and almost constant IBR. Now the main focus of attention is turned to the evolving peak asymmetry, and the crossover from micro- to macroplastic deformation.

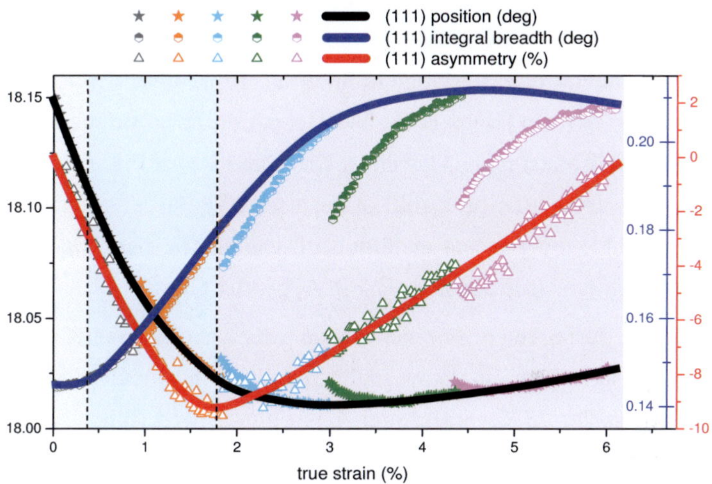

Figure 5.13.: Evolution of peak position, IBR, and asymmetry for the loading segments of the five deformation cycles. The solid lines represent the envelopes from Fig. 5.1. The deviation from linear behavior of peak position and increasing IBR indicate the crossover from elastic to microplastic deformation. The reversal of peak asymmetry ($\varepsilon \approx$ 1.8%) manifests the transition from microplasticity to macroplasticity.

The microplastic regime is defined as a transitional regime between purely elastic and macroplastic deformation, and ranges over a widespread strain regime for NC metals [Saada, 2005; Brandstetter et al., 2006]. In order to characterize this microplastic regime it is first referred to the lattice strain evolution of different (hkl) planes (see Fig. 5.14(a)). According to their elastic compliance, (200) oriented grains bear more elastic strain than (220), than (111) leading to very different strains within individual grains of the NC aggregate. This heterogeneous response enforces accommodation processes. However, due to the geometrical confinement, high stresses - which are not present at low applied strains - would be required to activate dislocation sources at GBs [Derlet et al., 2003b; Bachurin and Gumbsch, 2010] or within grains. Instead, in order to ensure overall compatible deformation, accommodation processes more likely occur in the GBs, e.g. in form of shuffling processes [Derlet et al., 2003b] or GB slip [Weissmueller et al., 2011]. In addition, triple lines acting as steric hindrance for GB slip [Hahn et al., 1997], restrict the accommodation and yield local back stresses on adjacent grains (see inset of Fig. 5.14(a)), and consequently, the elastic peak shift of the affected grains is restricted. This explains a (hkl)-independent evolution of asymmetric peak shape in the diffraction patterns (see also Fig. C.4 in Appendix C).

Generally, during the *in situ* mechanical tests it is consistently observed, that the peak asymmetry is always inclined towards the direction of the peak shift: For tensile, 2θ decreases and A is right-skewed (thin film chapter 5); for compression, 2θ increases and A is left-skewed (bulk chapter 4). A similar peak asymmetry evolved in diamond under hydrostatic loading conditions [Weidner et al., 1994]: As long as no macroscopic yielding was observed (high pressure, low temperature), a pronounced asymmetry evolved, due to heterogeneities in the diamond aggregate. However, when the diamond became ductile and yielded (high pressure, high temperature), the peak shape became symmetric again. A strain gradient along film thickness [Genzel, 1997] is excluded as a reason for peak asymmetry, as comparable peak asymmetries are observed for inert-gas condensed bulk samples with thicknesses in the mm range (cf. section 4.2.3). Planar faults could also cause an asymmetry [Balogh et al., 2006]. However, only individual (hkl) families would be affected, which is not the case in this study, where all reflections behave similarly (cf. Fig. C.4 in the appendix). Furthermore, twin dominated behavior would yield strongest broadening

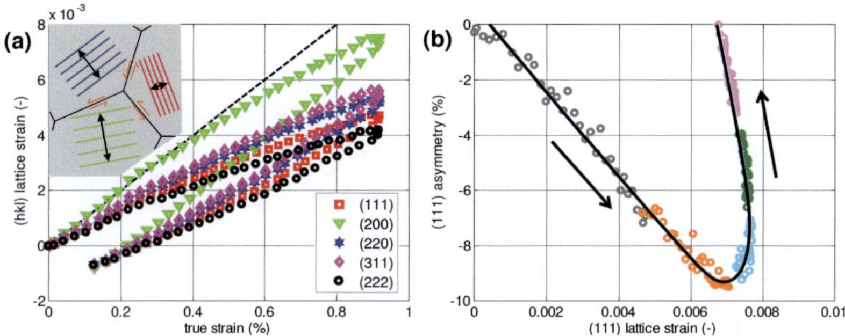

Figure 5.14.: (a) The evolution of elastic (hkl) lattice strains with ε demonstrates the varying lattice response as a result of differently compliant crystal orientations. In accordance with the elastic compliance, (200) planes bear more elastic strain than (220), than (111). Please note that the difference between (111) and (222) lattice strain is owed to the transmission geometry (cf. section 3.4.1 and Fig. 3.1(b)). Hindered accommodation processes in GBs and triple lines yield to back stresses and limit the elastic deformation within the affected grains (see inset). This accounts for the observed peak asymmetry. (b) When overall elastic deformation saturates, the asymmetry reverses as a result of macroplastic dislocation-based deformation.

increases for (h00) reflections [Brandstetter et al., 2008]. However, the (hkl)-dependent peak broadening of the tested Pd films, shown in Fig. C.6 in the appendix, is similar to a behavior dominated by elastic anisotropy (least increase for (hhh) and (h00) [Singh and Balasingh, 2001]) and dislocation-based mechanisms [Brandstetter et al., 2008].

Recently, it was shown that the GB accommodation processes may have an elastic and plastic nature [Weissmueller et al., 2011]. We argue that up to $\varepsilon = 0.3\%$, the accommodation is in principle of elastic nature, since IBR is fairly constant and peak position changes linearly. Subsequently, the continuously increasing asymmetry goes along with severely increasing IBR and increasing non-linearity of peak position, which are all indicators for non-elastic deformation. The severe broadening increases arise from the widespread distribution of grain orientation, size and corresponding lattice strain within the NC aggregate, providing diverse grain-to-grain interactions. Actually, this broadening increase is more pronounced than the subsequent broadening during macroplastic deformation.

The amount of peak asymmetry reaches its maximum at $\varepsilon = 1.8\%$. With further straining, the asymmetry reverses and the peak becomes more symmetric again. This maximum peak asymmetry denotes the beginning of the macroplastic regime. In Fig. 5.14(b), it is shown that the reversal of peak asymmetry occurs, when the overall elastic lattice strain saturates. With the release of the individual stress concentrations at triple lines, the deformation becomes more and more homogeneous. With increasing ε, the overall stress level becomes high enough that an increasing number of grains can deform by dislocation plasticity. Analog to the hydrostatic loading conditions in Ref. [Weidner et al., 1994], the geometrical constraint of the NC microstructure retards dislocation activity. The evolving peak asymmetry has been explained in both cases as a result of the hindered accommodation of the extraordinary large and differing lattice strains. However, if dislocations can be (nucleated and/or) activated, peak asymmetry diminishes in both cases. In addition to the current understanding [Saada, 2005; Brandstetter et al., 2006], it is argued that, not only the spread in the elastic limits based on the grain size distribution, but the heterogeneous elastic response and the resultant complexity of its accommodation, account for the extension of the microplastic regime in NC metals.

Making use of the in-plane information revealed by an area detector, the comparison of the evolution of peak asymmetry in loading direction ($\phi = 90°$) over ε with the in-plane evolution of the INT (displayed for ϕ directions of INT maxima and minima), impressively brings out the concurrence of reversing asymmetry and onset of in-plane texture evolution in form of a six-fold symmetry (see Fig. 5.15). This type of preferred orientation with (111) planes dominant in tensile direction is typical for tensile deformed FCC CG metals, particularly for wires and rods [Barrett and Massalski, 1966; Gambin, 2001]. The onset of texture formation is evidence of prevailing dislocation plasticity and, besides the reversing asymmetry, another striking argument for the strain-dependent crossover from micro- to macroplasticity in the tested Pd thin films ($\varepsilon_{x-over} = 1.8\%$). The explicit concurrence of onset of texture formation and asymmetry reversal provides strong evidence that the reversal of peak asymmetry, resulting from more homogeneous deformation, is owed to intragranular dislocation plasticity and not from stress release in the GB network, e.g. resulting from GB shear, slip, or migration. The sharpness of texture mainly depends on two factors: (i) The degree of deformation and (ii) to what extent dislocations contribute to overall deformation. An attempt to compare the contribution of dislocation plasticity to overall deformation between thin films and the various bulk samples is made in chapter 6 paying special attention to differences in grain size and fabrication process. In summary, three individual peak parameters are identified, which pinpoint the onset of dislocation-based plasticity: (i) Saturation of lattice strain / peak shift, (ii) reversing peak asymmetry, and (iii) onset of in-plane intensity redistribution.

Stress-driven GB migration was recently identified as a relevant deformation mechanism in NC metals [Rupert et al., 2009; Gianola et al., 2006]. The contribution and occurrence of this mechanism is strongly affected by the impurity content as GB mobility can diminish with increasing impurities, e.g. as a consequence of higher base pressure during fabrication [Gianola et al., 2008; Thompson, 1993]. Compared to the samples of Refs. [Gianola et al., 2006, 2008], the base pressure during Pd film fabrication was almost one order of magnitude lower. Furthermore, noble Pd is much less prone to oxidation than Al. Both facts may account for the low resistance to grain growth of the herein tested NC Pd, resulting

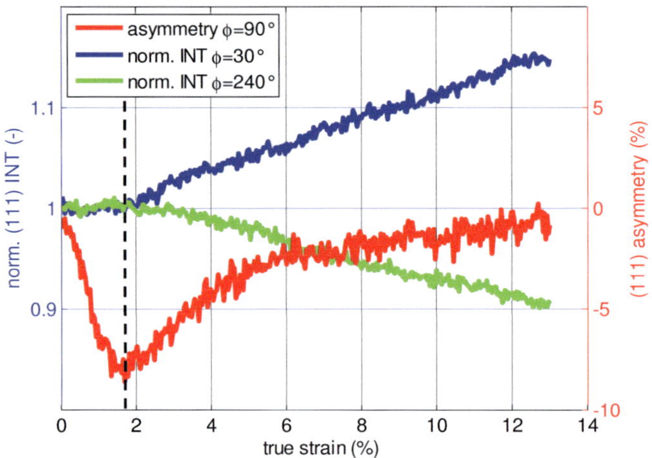

Figure 5.15.: Masterplot combining the evolution of peak asymmetry and in-plane INT redistribution over ε. Onset of texture formation coincides with that strain value, where the asymmetry reverses ($\varepsilon_{x-over} = 1.8\%$).

in grain growth even for $\varepsilon \leq 0.9\%$ and significant contributions of GB migration to overall deformation.

Although grains grow, no reduction in microstrain was measured, as one would expect from a $D \propto 1/<\varepsilon>$ scaling [Ames et al., 2008]. Instead, $<\varepsilon>$ increases from cycle to cycle, with a trend similar to results from virtual diffraction on NC Pd [Markmann et al., 2010]. These results and complementary MD simulations [Bachurin and Gumbsch, 2010] suggest that microstrain can increase significantly, even in the elastic regime and although dislocations are absent for $\varepsilon < 3\%$. It is argued, that the increase in $<\varepsilon>$ rather results from the elastic inhomogeneous response than from increasing dislocation density / activity, which agrees with the explanation given above.

In summary, it was demonstrated that the peak asymmetry and the in-plane intensity distribution are sensitive parameters to draw the line between micro- and macroplastic deformation of NC Pd. As long as dislocation activity is retarded, a broad distribution of elastic strains between individual grains and the complexity of their accommodation in GBs and triple lines cause a distinct peak asymmetry. Once dislocation plasticity can be activated for an increasing number of grains at higher strains, the strain distribution becomes more homogeneous, leading to a reversal of peak asymmetry. Simultaneously, a six-fold symmetry of INT along ϕ evolves similar to tensile deformation textures of CG FCC metals [Gambin, 2001], indicating that intragranular dislocation plasticity is responsible for the reversal of peak asymmetry. Independently, stress-driven GB migration is active, even at $\varepsilon \leq 0.9\%$. After $\varepsilon_{max} = 6\%$, the average grain size increased by 18% from 28 nm to 33 nm. The significant grain size increase implies that GB migration considerably contributes to overall deformation. A comparison of the relative contributions of individual deformation mechanisms with the bulk samples from chapter 4 is given in the comprehensive discussion in chapter 6. Overall, the careful analysis of *in situ* XRD data has demonstrated that following the evolution of peak parameters, especially of peak asymmetry and integral peak intensity, gives additional insight in the complex interplay of GB- and dislocation-based deformation mechanisms of NC metal thin films.

5.3.2. Alloying Effects in PdAu Thin Films

The principle deformation behavior of NC thin films and the classification into different regimes on the basis of differing dominant deformation modes has been treated in the former section. In the following, differences in the deformation behavior of PdAu alloys with Au contents ranging from 12 at% to 72 at% are discussed and compared to the behavior of pure Pd. But first, the initial state of the different alloys and the pure Pd is discussed.

5.3.2.1. Alloying Effects on the Initial State

A linear relationship between lattice constant a and alloy composition is manifested for the investigated alloys. The lattice constant a is calculated from the initial $2\theta_{111}$-angle of each undeformed sample and the Au content of each sample is determined by energy-dispersive X-ray spectroscopy (EDX) measurements. The result is shown in Fig. 5.16 and yields excellent agreement with the predictions of Vegard's Law [Vegard, 1921]: (i) The calculated a value of pure Pd matches the theoretical value, (ii) the alloys show a linear relationship of a over Au content, and (iii) even the prediction for pure Au based on a linear extrapolation of the computed a values over Au contents merges to the theoretical value. However, it is pointed out that the measured a values are not strain-free and deviations may arise from slightly different residual stresses for the different alloy compositions as a result of film deposition by magnetron sputtering.

In the following, the differences in the initial microstructures are discussed. Fig. 5.7(c) displays largest initial peak broadening for intermediate alloy compositions. Indeed, peak broadening analysis by SLM (Fig. 5.9) yields clearly largest grain sizes ($D = 30\,\text{nm}$) and lowest microstrain for pure Pd. All alloys exhibit smaller grain sizes, ranging from $20\,\text{nm}$ to $24\,\text{nm}$ without a visible alloying trend. However, the following tendency becomes apparent: The purer the material, the larger the grains and the lower the microstrain, which is known from literature, e.g. so-called solute drag [Koch et al., 2008; Wang et al., 2007] as for the alloys in this case, or impurity drag [Gianola et al., 2008]. Vice versa, the microstrain is considerably enhanced for $12\,\text{at}\% < c_{Au} < 29\,\text{at}\%$, compared to pure Pd, while for $53\,\text{at}\% < c_{Au} < 72\,\text{at}\%$ only moderate increases are measured. If the solutes would segregate to the GBs, a reduction in GB energy could possibly be identified by reduced microstrain. However, the opposite is observed: Alloys with intermediate solute contents exhibit highest initial

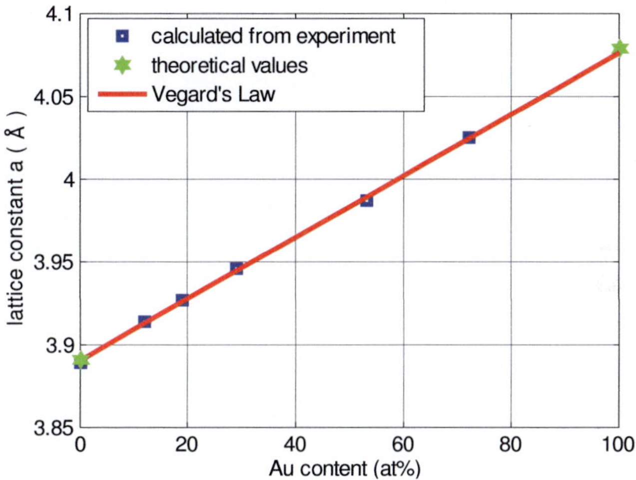

Figure 5.16.: Alloy-dependent lattice constant *a* as a function of Au content. The exact Au contents of the alloys were measured by EDX (see section 3.5.3).The linear relationship in a continuously miscible ally system as predicted by Vegard's Law is fulfilled. Theoretical values are from Ref. [Beck, 1995].

microstrains. Therefore, GB segregation is rather excluded, and the effect of solutes on the microstructure is rather related to pinning of GBs (kinetic approach) than to a reduction of GB energy (thermodynamic approach) [Koch et al., 2008].

TEM analyses of the PdAu alloys [Castrup, 2012] reveals a much higher density of growth twins for alloys with higher Au contents ($c_{Au} > 50$ at%). Single grains contain several stacked twins, with the coherent twin boundaries (TB) preferentially oriented parallel to the sample surface. The enhanced twin density could explain the observation of fairly constant coherent scattering domain size among all alloys, though reduced $<\varepsilon>$ for highest Au contents, since the coherent TBs are energetically more favorable than GBs, bringing the system closer to equilibrium. Note that in this *in situ* geometry, not the spacing of coherent TBs is probed as coherent scattering domains, but the column diameter. Therefore, owing to the transmission geometry, the effect of the stacked twins cannot be reflected by the probed domain size, however the sharp TB cause a $<\varepsilon>$ reduction also in the direction probed by XRD.

Initial fiber texture in growth direction, resulting from the deposition process, might also influence the deformation behavior. To quickly assess the sharpness of the $<111>$ fiber texture, which is often prevalent in sputter-deposited FCC thin films, the initial (111) peak amplitudes are divided by the (200) peak amplitudes (cf. Fig. 5.6), and shown in Table 5.1. The alloyed films clearly exhibit indications for stronger fiber texture compared to pure Pd, which shows the lowest calculated ratio of amplitudes. Alloys ranging from 12 at% $< c_{Au} <$ 29 at% exhibit the highest ratios , while for higher Au contents ($c_{Au} >$ 50 at%), the ratio decreases again. Nevertheless, all values are significantly higher than the theoretical values for pure Pd and pure Au, based on calculations employing the form factor and other affecting XRD parameters. This means, a fiber texture exists in all investigated thin films, with the trend that intermediate alloy compositions (12 at% $< c_{Au} <$ 29 at%) exhibit strongest texture.

Table 5.1.: Characterization of sharpness of the initial fiber-texture by relating the initial (111) to (200) peak amplitudes. The theoretical values are based on calculations employing the form factor and other XRD factors.

sample	Pd	Pd-12Au	Pd-19Au	Pd-29Au
$A_{0,111}/A_{0,200}$	3.66	4.56	4.74	4.57

sample	Pd-53Au	Pd-72Au	Pd_{theo}	Au_{theo}
$A_{0,111}/A_{0,200}$	4.45	4.18	2.02	1.95

5.3.2.2. Alloying Effects on the Deformation Behavior

In contrast to the peak position, the elastic lattice strain (Fig. 5.7(b)) is unaffected by the different a values, since it is calculated by relating the actual Δd to the initial d_0 value (see Eq. (3.3)). One reason for the generally higher lattice strains of the alloys compared to pure Pd is certainly the different grain size, as seen in Fig. 5.9: Pure Pd exhibits by far a larger grain size and therefore, assuming classical Hall-Petch behavior for this grain size regime, bears less lattice strain than the alloys with smaller grain size. A solid solution strengthening effect is expected to be rather small, as the difference in atomic radii is less than 4% [Labusch, 1970] ($r_{Pd} = 140\,\mathrm{pm}$ and $r_{Au} = 135\,\mathrm{pm}$ [Slater, 1964]). Among the different alloys, alloys with high Au content bear more lattice strain than samples with lower Au content. This difference may arise, since the microstructures of the alloys with high Au content contain more growth twins. This may provide an additional reduction of the average internal dimension responsible for Hall-Petch like strengthening, which is not necessarily seen in the coherent scattering domain size extracted with this *in situ* setup geometry. In literature it is argued, that besides GBs also twin boundaries serve as barriers for dislocation motion, leading likewise to a $\sigma_y \propto d^{-1/2}$ scaling law [Shen et al., 2005; Lu et al., 2009] and hence may account for the even more increased lattice strain for alloys with high Au contents.

After $\varepsilon_{max} \approx 1.8\%$ all tested samples exhibit larger grains and enhanced microstrain. In contrast to the alloys, the microstructure of pure Pd is lacking of solutes, which could retard GB migration and hence slow down grain growth. As a result the pure Pd exhibits a slightly larger grain size increase. On the other hand, the samples with the lowest initial microstrain, show strongest $<\varepsilon>$ increase. As it was argued in the former section, the increase of $<\varepsilon>$ in the microplastic regime ($\varepsilon < 1.8\%$) is rather attributed to the elastic inhomogeneous response than to increased dislocation density / activity. Based on this and the fact that during loading these alloys show the most pronounced increase in normalized IBR (see Fig. 5.7(d)), one may speculate that the alloys with low Au contents (12 at% $< c_{Au} < 29$ at%) deform more homogeneous than the pure Pd and the alloys with higher Au contents. This idea is supported by the fact that the low Au alloys show the most pronounced initial <111> fiber texture. The more directional microstructure can deform elastically more homogeneously compared to a more randomly distributed structure, where elastic anisotropy leads to stronger grain-to-grain interactions, and finally to more complex and also to irreversible accommodation processes. This hypothesis is verified by plotting both, the initial ratio of (111) to (200) amplitudes (cf. Table 5.1) and the deformation-induced absolute change of asymmetry ΔA over Au content (cf. Fig. 5.8), as shown in Fig. 5.17. The ratio of peak amplitudes is indicative for the sharpness of the initial fiber texture and the absolute change of peak asymmetry is a measure for the heterogeneity of deformation in the microplastic regime, as it was argued in section 5.3.1. It is in evidence, that the stronger the initial fiber texture is, the more homogeneous deformation proceeds (in the microplastic regime), which was deduced from the diminished change in asymmety, ΔA. The similar trend is observed for different AuCu alloy compositions in thin film geometry (cf. inset of Fig. D.2(c)).

The strain increase tests of pure Pd (Fig. 5.1) and Pd-12Au (Fig. 5.10) are opposed in Fig. 5.18, showing the envelopes of the loading cycles of different (111) peak parameters. The alloyed sample is discussed as being representative for the alloy compositions with lower Au contents (12 at% $< c_{Au} < 29$ at%). At first, the higher lattice strain values (dashed lines) of the alloy are mentioned. As discussed above, this is mainly attributed to Hall-Petch strengthening, since the alloy measures a smaller initial grain size (see Fig. 5.19(a)). Furthermore, since the evolving lattice strain is only a measurement relative to the initial state, different residual stresses might also entail a difference in the measured values. Analysis of the fabrication process of these films [Castrup et al., 2011] shows, that depositing pure Pd and Pd-10Au at the same Ar pressure (e.g. 5×10^{-3} mbar) results in stronger compressive stresses for the alloy. As

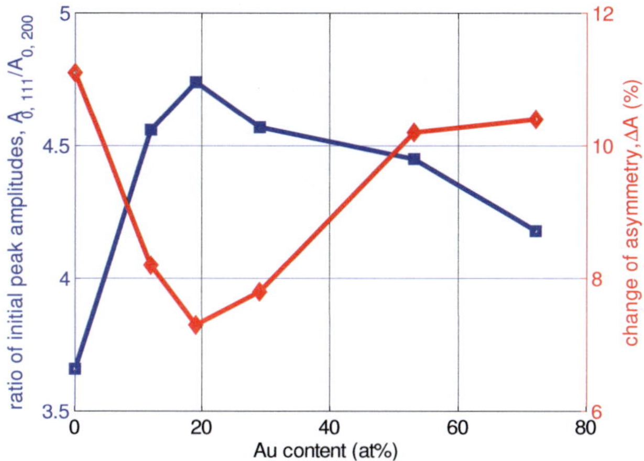

Figure 5.17.: Ratio of initial (111) to (200) peak amplitudes, as an indicator for the sharpness of the fiber texture and the absolute change of peak asymmetry ΔA at $\varepsilon = 1.7\%$, as a measure for the inhomogeneity of microplastic deformation, are correlated as a function of the Au content.

a consequence, the displayed difference in elastic lattice strain may be overestimated. However, unaffected by possibly different residual stress states is the steeper linear lattice strain increase for the alloy, which was also found in Fig. 5.7(b) for other alloy compositions. This is surprising, since the Young's Modulus should be constant in this alloy range, as it was shown in Fig. 4.23. The stronger deviation of pure Pd from the ideal elastic slope of 1 thus may indicate that more deformation is accommodated in or close to the GBs and not in the grain interior which is represented by the lattice strain.

Seizing the reverse behavior of enhanced initial fiber texture and reduced asymmetry evolution, discussed in Fig. 5.17, the reduced increase of normalized IBR of the alloy during loading, shown in Fig. 5.18 with dotted lines, is also attributed to the more homogeneous microplastic deformation,

133

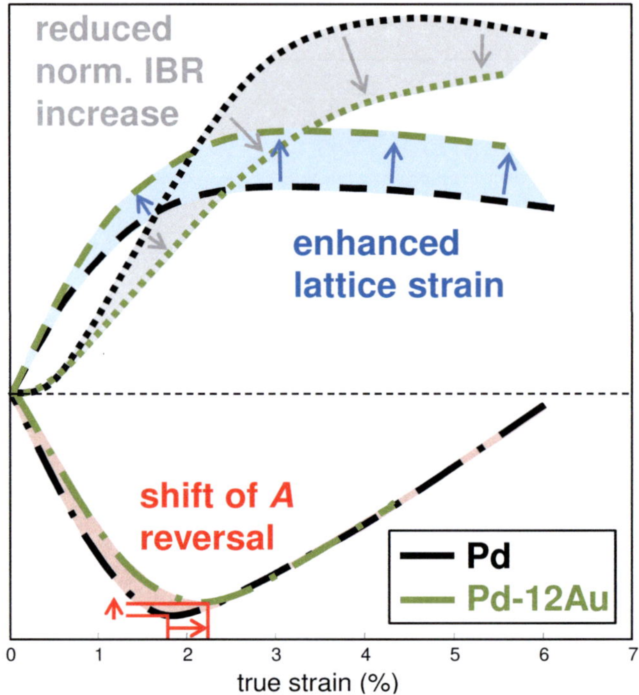

Figure 5.18.: Masterplot comparing (111) parameters of the strain increase tests of Pd and Pd-12Au. Solely the envelopes of the loading cycles are shown. The dashed lines represent the lattice strain, the dotted lines the normalized IBR and the dashed-dotted lines the asymmetry.

with less grain-to-grain interaction resulting from the more pronounced $<111>$ fiber texture. When grains are oriented in the same crystallographic orientation, the effect of elastic anisotropy is reduced, compared to an isotropic polycrystal. This also explains the reduced absolute change of asymmetry before the reversal point (dash-dotted lines). Furthermore, the crossover strain ε_{x-over}, where the asymmetry reverses, is located at a higher value $\varepsilon_{x-over} = 2.2\%$ for Pd-12Au compared to $\varepsilon_{x-over} = 1.8\%$ for pure Pd. Consequently, it is reasoned that alloying causes an extension of the microplastic regime, and hence a further retardation of dislocation-based plasticity to higher strains. On the other hand, the alloy seems to deform more homogeneous in the extended microplastic regime, indicated by reduced normalized increase of IBR and reduced absolute change of peak asymmetry.

The reduced grain-to-grain interaction as a result of increased initial fiber texture and the less pronounced and deferred dislocation activity, yield a lower increase of $<\varepsilon>$ over ε_{max} for the alloyed sample, as seen in Fig. 5.19(b). Finally, the stress-induced grain growth is discussed (Fig. 5.19(a)). Starting from initially different grain sizes, pure Pd and Pd-12Au exhibit fairly the same increase in grain size over the applied strain. As mentioned before, the small difference in atomic radii (less than 4%), seems to be rather ineffective in retarding GB migration.

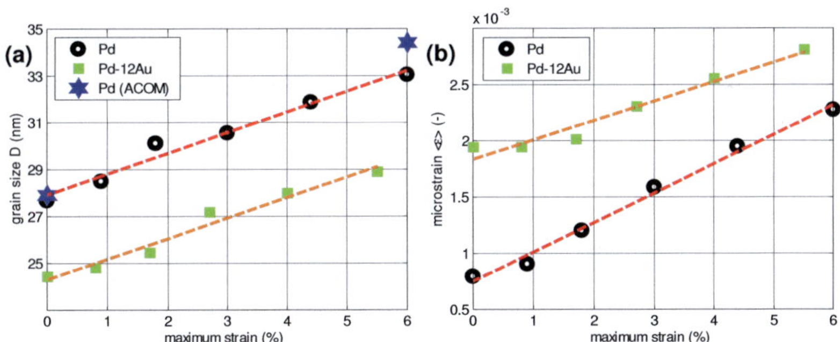

Figure 5.19.: Comparison of grain size and microstrain, calculated by the SLM, for pure Pd and Pd-12Au in the unloaded states after individual loading cycles.

Similar to pure Pd (Fig. 5.5), prevailing plastic deformation yields an emerging shear band pattern as seen in Fig. 5.12(c) for the Pd-12Au alloy. Since the plastic deformation localizes in these shear bands, and NC metals generally lack in strain hardening, as indicated by low irreversible peak broadening shares after unloading (cf. [Budrovic et al., 2005]), the shear bands are precursors for film delamination and cracking [Lohmiller et al., 2010]. Indeed, the onset of film cracking for the Pd-12Au sample is identified at $\varepsilon \approx 5\%$, which was approved by *in situ* tensile testing under an optical microscope, while pure Pd remained free of cracks, even after $\varepsilon_{max} = 13\%$. Two possible reasons for the reduced ductility are noted: (i) The ability for strain hardening is reduced in the alloy, e.g. due to the smaller grains, and the deferral of dislocation-based plasticity to higher strains, and (ii) the adhesion of the film to the substrate could be reduced by the Au solutes, as pure Au tends to form islands during initial deposition.

To conclude this section, the classification into different deformation regimes, introduced in Fig. 5.13 for pure Pd, is basically equivalently valid for the PdAu alloys. No drastic change from one prevailing mechanism to a fundamentally different mechanism was observed, since there is not one single governing deformation mechanism, but always an interplay and intermixing of different mechanisms. Rather the relative contributions from individual mechanisms to overall deformation can be influenced by alloying. By adding Au to Pd, initially more deformation is accommodated in the grain interior. As a result of the sharper fiber texture, measured for the alloys, their deformation is more homogeneous in the microplastic regime: The effect of elastically differently strained grains, leading to inhomogeneities, is reduced. For alloyed samples with less local stress concentrations and with generally smaller grain sizes, the macroplastic regime, governed by dislocation plasticity, is deferred to higher strains. On the other hand, all alloys show a reduced ductility compared to pure Pd, which is deduced from earlier film cracking. A stronger effect on the deformation behavior might occur for alloys with very high Au contents, where more growth twins could have a distinct influence. However, in that case, alloying and microstructural effects would correlate and the separation would be even more delicate.

5.4. Summary

The mechanical behavior of polyimide-supported NC Pd thin films, revealed by synchrotron-based *in situ* tensile testing and sophisticated peak shape analysis, can be summarized as follows:

- Based on the evolution of several XRD peak parameters, different deformation modes are identified, separated, and classified to specific strain regimes.

- Initially, heterogeneous lattice elasticity is dominant, accompanied by accommodation processes in GBs and triple lines.

- With increasing strain, the accommodation among the differently strained crystals is restricted at GBs and triple lines and yields local stress concentrations.

- Onset of dislocation-based plasticity is observed at strains $\varepsilon >$ 1.8%, which was deduced from the evolution of several XRD peak parameters individually (position, asymmetry, intensity distribution).

- Stress-driven GB migration is independently active, even for $\varepsilon < 0.9\%$, and accounts for considerable contributions to overall deformation.

Alloying a limited amount of Au ($c_{Au} <\approx 30$ at%) to Pd has the following effects:

- Distinct reduction of initial grain size D, although the degree of grain growth during deformation is rather unaffected.

- Due to stronger initial fiber texture, microplastic deformation is more homogeneous involving eased grain-to-grain interactions.

- The microplastic regime is extended to higher strains and likewise the onset of dislocation-based plasticity is further deferred.

- The ductility is clearly reduced, leading to earlier film cracking.

For alloys with $c_{Au} >\approx 50$ at%, a modified initial microstructure is observed incorporating more growth twins, which complicates the untangling of alloying effects and microstructural effects.

6. Comprehensive Discussion and Outlook

The aim of this thesis has been to systematically investigate deformation mechanisms emerging in NC metals and alloys during mechanical testing. This goal was pursued by applying synchrotron-based *in situ* mechanical testing to different NC metals and alloys. Therefore, different setups, adapted for the demands of the different sample types and geometries, were developed and/or improved. Three major experimental and methodological novelties compared to earlier work were established in this thesis:

1. In addition to thin film tensile testing, NC bulk metals and alloys have been tested under compressive and shear-compressive loading using a setup comprised of a high energetic and microfocused X-ray beam (High Energy Microdiffraction (HEMD) endstation at the ESRF).

2. Data evaluation was based on sophisticated peak shape analysis involving several important peak parameters, including position, breadth, intensity, and asymmetry, which help to unravel different deformation modes. Subsequent line broadening analysis allowed to monitor grain size and microstrain during mechanical testing.

3. The potential of area detectors was exploited, analyzing several complete Debye-Scherrer rings by radial scans with an azimuthal increment of $\Delta\phi = 2°$. Thereby, formation of deformation textures could be monitored qualitatively, as well as, the evolution of in-plane grain shape. Both features contributed significantly to identify and untangle coexisting mechanisms.

In the following, the main insights from bulk compression testing and thin film tensile testing of the different materials are reconciled. The aim is to excerpt principles of how NC metals deform in general and to carve out pronounced differences between the different material systems. In order to focus on the main aspects, a comparison is drawn between Pd (thin film), Ni (bulk), and Pd-30Au (bulk) as representative for the IGC-fabricated PdAu alloys. For the sake of clarity, only one representative for each material system is discussed. Alloying effects of IGC-fabricated and sputter-deposited PdAu alloys are discussed in detail in section 6.3. The main characteristics of the samples discussed in the following are summarized in Table 6.1.

Table 6.1.: Main characteristics of the three different material systems.

Sample	Ni	Pd-30Au	Pd
Processing	PED	IGC	RF sputtering
Sample geometry	bulk	bulk	thin film
Initial grain size (nm)	30	10	28
Loading condition	C	C	T

6.1. Elastic Lattice Strain and Asymmetry

The discussion starts with the comparison of the evolution of lattice strain and peak asymmetry. In order to ease the comparison of the different loading conditions, the absolute values of lattice strain and asymmetry are used. The polar plots of Fig. 4.17 have evidenced that a negative 2θ-shift causes a right-skewed asymmetry for tensile loads and vice versa, a positive 2θ-shift causes a left-skewed asymmetry for compressive loads. The tensile tested thin films revealed the similar finding (Figs. 5.1 and 5.3). While the leading flank of the peak accompanies the overall shift, the trailing flank lags behind, so entailing an asymmetric peak shape. The conversion of the pristine asymmetry parameter ensures equal absolute values for both types of asymmetry (cf. section 3.3.2). Therefore, the results from bulk compression testing and thin film tensile testing can be directly compared. The consistent behavior of interacting peak position and peak asymmetry is visualized in Fig. 6.1.

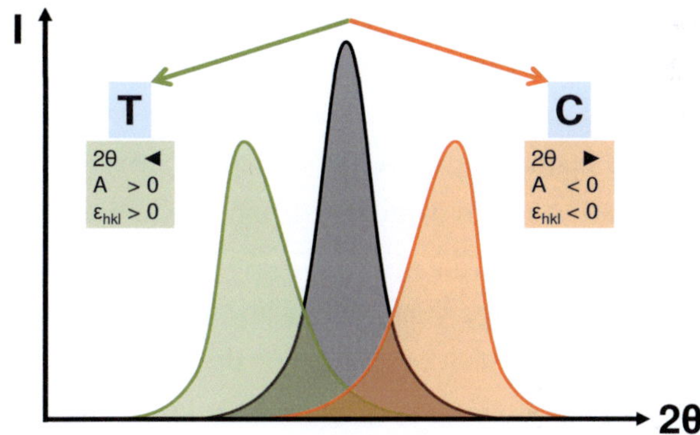

Figure 6.1.: Generalized behavior correlating the directions of peak shift and asymmetry. For the sake of simplicity, the schematic neglects texture formation, which would result in (hkl)-dependent changes of the integral peak intensity: Dependent on the considered (hkl) plane, e.g. for (111) or (200), tensile (T) would cause an INT increase and compression (C) a decrease **or** e.g. for (220) the opposite behavior.

The evolution of (111) lattice strain and asymmetry in loading direction ($\phi = 90°$) for the three different material types is shown in Fig. 6.2. At first, the generally high elastic lattice strains inherent to NC metals are mentioned. Owing to the small grain sizes, NC metals possess high strengths compared to CG counterparts. The fact that the Young's Modulus is a material constant and should not be affected by the microstructure[1] implies that the high yield strength of NC metals is directly related to high

---

[1] Indeed, residual porosity or a textured microstructure could change the apparent Young's modulus.

elastic strains, leading to an extended elasticity-dominated regime. In Fig. 6.2(a), the absolute values of the evolution of (111) elastic lattice strain in loading direction ($\phi = 90°$) are displayed for the three different NC materials. During initial loading, the materials show similar lattice strain evolution. However, for higher strains, major differences, dependent on the nature of plastic deformation, are observed. For Ni and Pd, with grain sizes in the range of $D \approx 30\,\text{nm}$, the lattice strain saturates for higher total strains, or even decreases in the case of Pd as a result of shear localization (see Fig. 5.5). Nevertheless, in both cases it was argued that dislocation-mediated plasticity bears a major contribution to overall deformation. On the other hand, for the Pd-30Au alloy with $D \approx 10\,\text{nm}$, the lattice strain increases continuously during loading up to $\varepsilon = 28\%$. This illustrates the arduousness of the material to convert the externally applied work into dislocation-based deformation processes, which is attributed to the very small grain size.

The evolving peak asymmetry (Fig. 6.2(b)) yields further indications for the dissimilar behavior of the three materials. The reversal of the asymmetry in Pd, was interpreted as the transition from heterogeneous microplastic to dislocation-mediated macroplastic deformation (cf. section 5.3.1). The coincidental appearance of asymmetry reversal and onset of texture formation was also shown in Fig. 5.15.

Similar to dislocation plasticity, GB migration could also abate stress concentrations at GBs and triple lines, and thereby cause the reversal of peak asymmetry. However, for the IGC-fabricated PdAu alloys, GB migration was clearly identified, but peak asymmetry did not reverse (see section 4.3.3). Furthermore, no considerable deformation texture formed. Consequently, it is argued that the asymmetry reversal is correlated to texture formation, and hence triggered by intragranular dislocation plasticity, rather than by GB migration.

Although the absolute asymmetry is not that high for Ni, a distinct reversal is identified with the onset of texture formation. Similar to the

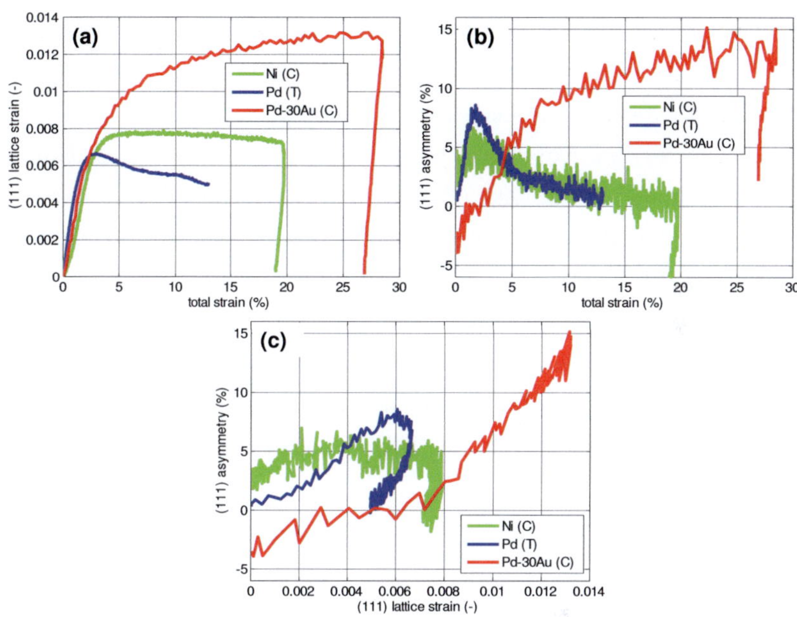

Figure 6.2.: Comparison of (111) XRD data in loading direction ($\phi = 90°$) for Ni, Pd, and Pd-30Au. (a) Lattice strain over total strain, (b) asymmetry over total strain, and (c) asymmetry over lattice strain. Regardless of compressive (C) or tensile (T) loading, as well as, left- or right-skewed asymmetry, the absolute values are compared. In (c) only the data for the loading sequence is shown.

argument in section 5.3.2, where the mitigation of deformation-induced changes of peak asymmetry, ΔA, from Pd to PdAu alloys, was correlated to sharper initial fiber texture (cf. Fig. 5.17), it is argued that the initial $<111>$ fiber texture of PED Ni in growth direction could have an equal effect, leading to mitigated ΔA of Ni compared to Pd, as a result of alleviated grain-to-grain interactions of the more directional microstructure. For highly fiber textured AuCu thin film alloys, a mitigated asymmetry change with stronger initial $<111>$ fiber texture is observed as well (cf. inset of Fig. D.2(c)). Again, the behavior of Pd-30Au is explicitly different, because the asymmetry does not reverse. Instead, similar to the lattice strain, the asymmetry continuously increases with further straining, as dislocation plasticity is still constrained.

In Fig. 6.2(c), the evolution of peak asymmetry is shown as a function of the lattice strain. For the sake of clarity, only the loading data of the tests are displayed. With this plot, it conclusively becomes apparent that as soon as the lattice strain saturates, the peak skews back toward symmetric peak shape (Ni and Pd, dislocation-mediated plasticity), while for continuously increasing elastic strains, the peak becomes more asymmetric (Pd-30Au, inhibited dislocation plasticity). Overall, the distinctly different behavior of Pd-30Au is attributed to the very small grain size, rather than to alloying effects.

6.2. Deformation Texture

In the following discussion of deformation textures the Ni SCS and the different IGC PdAu alloy compositions are included. The (111) texture ratio, relating the averaged (111) intensities of the six peak maxima to the six peak minima along ϕ at a given state of deformation, is employed to assess the individual contribution of dislocation plasticity for each sample type. In Fig. 6.3, the corresponding ratios calculated for $\varepsilon = 10\%$ and $\varepsilon = 20\%$ are compared. Clearly, Ni yields larger ratios for both loading conditions and both deformation states, when compared to Pd and PdAu thin film and bulk samples. Interestingly, the ratio for the SCS sample is much higher than for the COMP sample, alluding to the higher shear components promoting dislocation plasticity. For the IGC PdAu alloys, the texture ratio is generally much lower, owing to the inhibited dislocation plasticity for the very small grain size. However, a distinct alloy dependence is observed at $\varepsilon = 20\%$, with strongest values for highest Au contents. At $\varepsilon = 10\%$, the texture is not that pronounced in the alloys, so the trend first begins to appear at $\varepsilon = 20\%$. For Pd thin films, the ratio is only calculated for $\varepsilon = 10\%$ due to limited ductility. Astonishingly, an intermediate value ranked between Ni and the PdAu alloys is obtained, although the Pd thin film is in the same grain size range as bulk Ni. It is reasoned about other origins which could explain the different texture ratios: (i) Stress-induced GB migration is clearly more pronounced for Pd thin films compared to bulk Ni and may contribute considerably to overall deformation, reducing the necessary share of intragranular plasticity. In fact, the relative grain size increase, constituted by dividing the ϕ-averaged grain sizes after deformation to the initial grain size, yields 18% for Pd for $\varepsilon_{plastic} < 6\%$ and only 9% for Ni after $\varepsilon_{plastic} = 19\%$. One apparent reason for the reduced resistance of Pd compared to Ni is the "cleaner" fabrication process. Sputter-deposition under conditions close to ultra high vacuum should lead to less impurities compared to the electrodeposition process involving plenty of additives in the bath. An enhanced impurity content can promote the pinning of GBs, and thereby impede GB mobility [Gianola et al., 2008; Tang et al., 2012]. (ii) The ratio of stable to unstable stacking fault energy is distinctly higher for Pd ($\gamma_{sf}/\gamma_{usf} \approx 0.84$ [Schaefer et al., 2011]) compared to Ni ($\gamma_{sf}/\gamma_{usf} \approx 0.55$-$0.70$ [Van Swygenhoven et al., 2004]). As a consequence, the intragranular plasticity of Pd is rather carried by full dislocations, while for Ni extended partial dislocations could play a major role [Van Swygenhoven et al., 2004]. Hence, cross-slip should be restricted in Ni, as only full dislocations can cross-slip [Bitzek et al., 2008], and thereby dislocations remain on their primary glide plane resulting in more pronounced texture formation in Ni.

All observed deformation textures correspond to typical texture formation found for CG FCC metals at high strains. Due to the technological relevance (e.g. rolling processes), the deformation textures of CG metals have been studied intensively [Wassermann, 1962; Barrett and Massalski, 1966; Gambin, 2001]. For compression (not rolling), the (110)

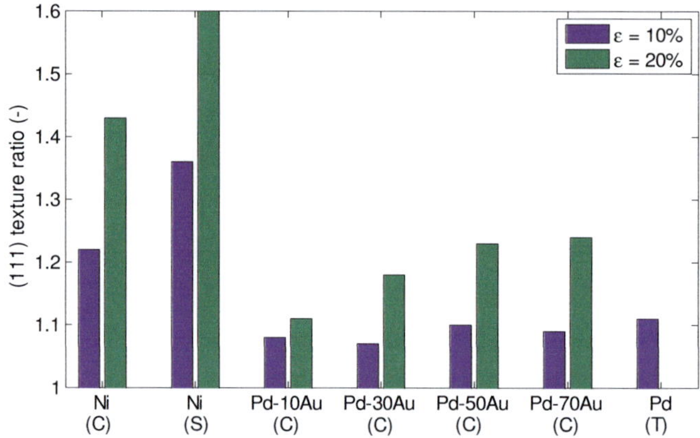

Figure 6.3.: The (111) texture ratio, relating the intensities of peak maxima to minima, calculated for all samples at $\varepsilon = 10\%$ and $\varepsilon = 20\%$. The loading conditions are indicated: (C) compression, (S) shear-compression, (T) tensile.

plane normal orients itself parallel to the compression direction, which is also observed for the tested bulk samples (see chapter 4). For SCS, the (110) planes orient likewise in the direction of maximal compressive strain, which is rotated by $\approx 7°$ with respect to pure compression. On the other hand, (111) and (100) plane normals orient in the direction of maximal tension which is the lateral direction, while their intensities are reduced in compressive direction. The tensile deformed thin films (see chapter 5 and Appendix D) exhibit a deformation-induced fiber texture similar to that of drawn wires, which is a duplex (111) + (100) texture. This means both (111) and (100) plane normals orient parallel to the tensile loading direction, while (110) planes align in the compressive lateral direction. Apparently, the tensile behavior is directly inverse to the compressive behavior. The overall behavior is summarized in Fig. 6.4, comprising the three different loading conditions: tensile (T), compressive (C) and shear-compressive (S). Regardless of the loading condition, the duplex

(111) + (100) texture always orients towards the maximum tensile direction (green arrows), while (110) rotates toward the maximum compressive direction (orange arrows).

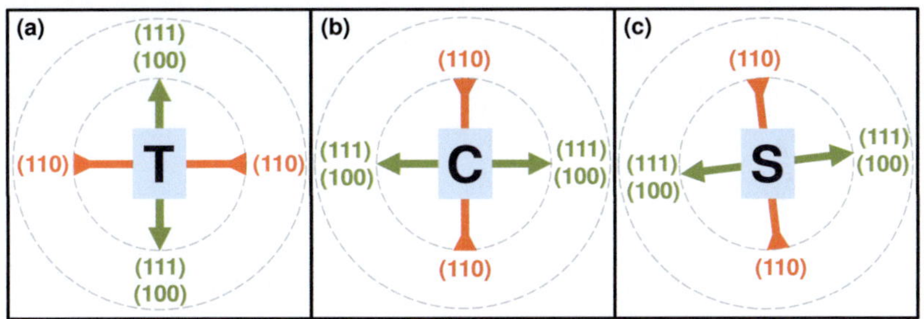

Figure 6.4.: Generalized orientations of the texture components with respect to loading condition. (a) Tensile (T), (b) compression (C), and (c) shear-compression (S). Regardless of the loading condition, the duplex (111) + (100) always orients to the maximum tensile direction, and (110) orients to the maximum compressive direction.

Finally it is noted that an initial fiber texture, as a consequence of the fabrication process, seems to be secondary. For PED Ni, it was shown by experiments with varying sample orientation that regardless of the initial in-plane INT distribution, the redistribution happens in the same way (Fig. 4.6). For sputter-deposited thin films, with an initial fiber texture in sample normal direction and random in-plane orientation, an in-plane texture with the (111) orientation dominant in tensile direction emerges and the initial out-of-plane texture attenuates.

6.3. Alloying Effects

Alloying effects for the continuously miscible PdAu alloy system were investigated with IGC bulk samples (cf. sections 4.2.3 and 4.3.3) and sputter-deposited polyimide-supported thin films (cf. sections 5.2.2 and

5.3.2). The data set is lacking in results on pure IGC Pd, since room temperature grain growth was observed [Ames et al., 2008], and therefore, it cannot be differentiated between thermally- and stress-induced changes of the microstructure. However, as a consequence, it is reasoned that by adding solutes the (initial) microstructure can be stabilized, which was also discussed in literature [Millett et al., 2007; Koch et al., 2008; Wang et al., 2007]. Similarly, for the thin films it was found that the adding of solutes can significantly minimize the initial grain size, compared to pure Pd. However, all alloys (bulk and films), as well as pure Pd thin films, have demonstrated that the microstructure is not stable upon mechanical testing, since all samples exhibit distinct grain size increases after deformation (Figs. 4.21, 5.9(a), and 5.19(a)). In fact, comparing pure Pd with Pd-12Au, both in thin film geometry (Fig. 5.19), it is observed that the ability of solutes to pin mobile GBs during loading is rather negligible. However, it must be noted that, the difference in atomic radii of Pd and Au ($r_{Pd} = 140\,$pm and $r_{Au} = 135\,$pm [Slater, 1964]) is rather low, which could be the reason for ineffective solute drag. Other solutes could be more effective in GB pinning, e.g. Pd-Ag alloys with $r_{Ag} = 160\,$pm [Slater, 1964]. In a very recent paper [Chookajorn et al., 2012], the theoretical framework was set to design stable NC alloys based on thermodynamic calculations, since until today solutes can only retard grain growth, but no complete prevention can be achieved [Weertman, 2012].

The impurity content is also known to have an impact on GB mobility [Gianola et al., 2008; Tang et al., 2012]. Assuming that the sputter-deposition process with a base pressure of $2 \times 10^{-8}\,$mbar is cleaner than inert-gas condensation ($2 - 5 \times 10^{-7}\,$mbar), it can be explained why the grain size of IGC PdAu increases only by 16% to 20% after 20% compressive strain (cf. Table 4.3), while for Pd and Pd-12Au films relative increases of 18% are measured already for less than 6% tensile strain (cf. Fig. 5.19(a)). Additionally, the residual porosity of the IGC samples

148

may have an influence and could aid with pinning GBs. The assumption of different impurity content seems to be plausible, because with equal amounts of impurities, IGC samples should exhibit more grain growth due to the smaller grain size and respectively larger GB volume fraction. However, the opposite behavior is observed.

Several effects seem to superimpose when comparing the mechanical behavior of pure Pd with Pd-12Au thin films. Besides effects on the intrinsic material parameters (e.g. stacking fault energy [Schaefer et al., 2011]), adding Au to Pd results in a decrease of initial grain size and an increase of the initial <111> fiber texture for the sputter-deposited thin films. Due to the limited ductility, the discussion emphasizes only the microplastic regime and the transition to macroplasticity. The enhanced lattice strain for the alloy is attributed to grain size strengthening, as Young's modulus does not change much in this alloy range (cf. Fig. 4.23). The effect of solid solution strengthening is expected to be small, as atomic radii are quite similar, and the difference in shear modulus, which can also account for solid solution strengthening [Labusch, 1970], is also rather small. The extension of the microplastic regime, and the concomitant deferral of dislocation-based plasticity can be partially attributed to the reduced grain size, but in addition, the stronger fiber texture reduces the magnitude of grain-to-grain interactions, and consequently stress concentrations at GBs and triple lines, which could initiate macroplastic dislocation activity. Thereby, an extended microplastic deformation is observed for the alloy (reduced change of peak asymmetry during loading, ΔA).

Alloying trends, over a broad composition range up to 70 at% Au, can be better discussed on the basis of IGC samples since for the sputter-deposited samples it was not possible to prepare comparable microstructures throughout the binary alloy system, e.g. thin films with large Au contents ($c_{Au} > 50$ at%) exhibit a significant increase in growth twins [Castrup, 2012], which strongly impact the mechanical response (cf. section 2.2.2). Due to the high strains which can be achieved by bulk compression experiments, the discussion of the IGC samples is focused on the macroplastic behavior. Alloys with high Au contents demonstrate more pronounced texture formation and stronger grain size increases. The first point is addressed to the reduced stacking fault energy [Schaefer et al., 2011], favoring extended partial dislocations [Van Swygenhoven et al., 2004] and reduced cross-slip probability [Bitzek et al., 2008]. The second point correlates with enhanced GB mobility, as a result of easier rearrangement processes near or at GBs due to the increased free excess volume. This trend was deduced from the ratio of background to peak intensity [Ungar et al., 2005, 2007; Lohmiller et al., 2012a], which increased most pronouncedly for alloys with high Au contents. On the other hand, Pd-10Au samples show crack formation after 7% compressive strain due to the contraints to dislocation plasticity.

Overall, compression testing of IGC PdAu alloys with $D \approx 10 - 15\,\mathrm{nm}$ involve comparable microstructures and high applied strains, and thereby it is found that with increasing Au content, dislocation-mediated deformation and GB migration become more important, while GB-mediated deformation becomes less important. The alloy-dependent behavior of thin films underlines this trend, where a decreasing deviation of lattice strain from the ideal elastic slope of 1 indicates reduced contributions of GB-mediated deformation (see Fig. 5.7(b)). The generalized alloying trends are displayed in Fig. 6.5.

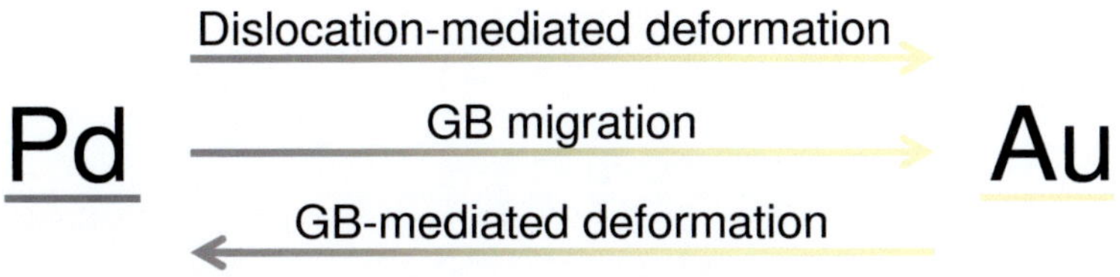

Figure 6.5.: Generalized alloying effects on the deformation mechanisms of the continuous miscible PdAu alloy system.

6.4. Succession of Deformation Mechanisms

Based on several independent XRD peak parameters, different deformation mechanisms were identified, separated, and classified in different strain regimes for all tested NC samples and loading conditions. A succession map (Fig. 6.6) was created in order to summarize the strain-dependent deformation behavior of the different material systems, and to be able to compare the relative contributions of individual mechanisms to overall deformation.

First, it must be noted that the Pd sample in the last row of the map is the only thin film sample which is discussed here, due to the limited ductility of the PdAu thin films. Second, the map should only give a qualitative impression of how the contributions of the individual deformation mechanisms vary between the different material systems, loading conditions, and alloying. The most important insights are summarized as follows:

- All tested samples show a succession and coexistence of several different deformation mechanisms, regardless of grain size, loading condition, or sample geometry.

- Although the relative contributions vary distinctly among the different samples, one common feature is that GB shear and slip is prevalent from the beginning, accompanying the elastic anisotropic

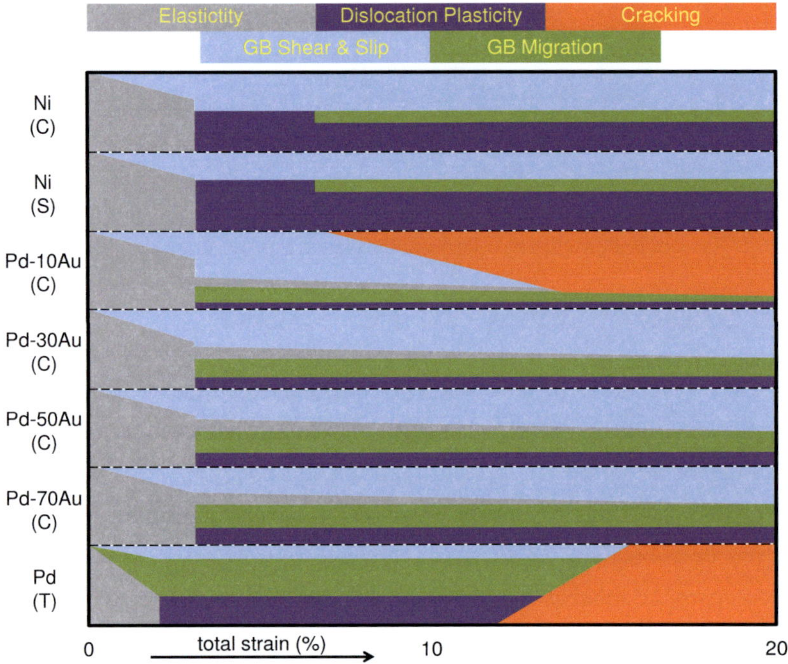

Figure 6.6.: Succession map displaying the succeeding and coexisting deformation mechanisms for the different material systems and loading conditions.

lattice deformation, which is inherently strongly pronounced for NC metals due to overall high lattice strains, owing to the constraints of ordinary dislocation plasticity.

- Dislocation plasticity is generally more pronounced in Ni and Pd with $D \approx 30\,\mathrm{nm}$ compared to PdAu alloys with $D \approx 10$ - $15\,\mathrm{nm}$. The geometrical confinement clearly aggravates dislocation processes. Possibly, the alloys are within the range of a strength-grain size interrelation, where strength reaches its maximum, as postulated in Fig. 2.1 and in Refs. [Meyers et al., 2006; Schiotz and Jacobsen, 2003; Trelewicz and Schuh, 2007].

- Shear-dominated deformation promotes dislocation plasticity in NC Ni and increases the relative contribution of dislocation-mediated plasticity.

- Among the discussed samples, significant differences with respect to relative contribution and onset strain of GB migration is identified. It is argued that the contribution and occurrence of GB migration strongly depends on impurities. Based on the obtained results it is speculated that sputter-deposition is "cleaner" than inert-gas condensation, than electrodeposition.

- Thereby, the enhanced contributions of dislocation plasticity and reduced contributions of GB migration for electrodeposited Ni compared to sputter-deposited Pd can be explained, although the grain size is similar. A decrease in the ratio of stable to unstable stacking fault energy from 0.84 for Pd [Schaefer et al., 2011] to 0.55-0.70 for Ni [Van Swygenhoven et al., 2004] would yield the same tendency and could additionally account for the described trend.

- The IGC PdAu alloys show clear trends to enhanced dislocation plasticity and enhanced GB migration for higher Au contents, while for Pd-10Au crack formation accompanies plastic deformation. In particular, the contributions from lattice elasticity to overall deformation are remarkable even at high plastic strains due to the very small grain size of 10 - 15 nm.

To further abstract the deformation behavior of NC metals it is focused on one predominant mechanism displayed as a function of grain size and degree of deformation, as shown in the deformation mechanism map of Fig. 6.7. The attempt is to embed the NC material behavior in-between conventional coarse-grained and amorphous material behavior.

By inspection of this rough simplification four points are noted: (i) In the NC regime, the strain range dominated by lattice elasticity increases with decreasing grain size. (ii) With decreasing grain size GB-mediated plasticity prevails over dislocation-mediated plasticity, which is illustrated in the map by a straight line as a first order approach. (iii) Pinning of GBs can significantly influence the interaction between the two deformation modes. (iv) The crossover from the NC to the amorphous regime is still not fully explored. The continuous reduction of grain size from 10 nm to the amorphous regime could shed light on this range, in addition to existing literature [Detor and Schuh, 2007; Trelewicz and Schuh, 2007].

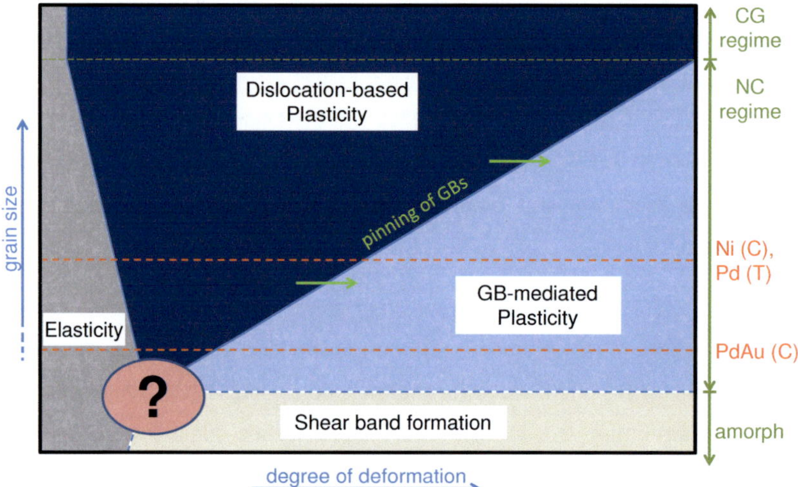

Figure 6.7.: Schematic deformation mechanism map denoting the prevailing deformation mechanism dependent on grain size and degree of deformation. The attempt tries to embed the NC material behavior in-between conventional coarse-grained and amorphous material behavior.

6.5. Outlook

Focusing research on one material system allows to vary individual microstructural parameters systematically. However, the behavior may be specific for the individual material system, and therefore the behavior derived from a single material system cannot be generalized for NC metals. In order to establish a generalized behavior for NC metals, different material systems must be characterized. For a successful comparison of the different systems, detailed knowledge of the underlying microstructure is a prerequisite to establish consistent microstructure-property relationships, and consequently to cross-link the different behavior of different systems. Related to this thesis, this means to further advance initial microstructural characterization. The characterization of impurity contents and segregations should especially be in the focus, since based on the results of this thesis it is speculated that this issue has a significant impact on the occurrence and contributions of GB-mediated plasticity and GB migration, and thereby on the interplay of dislocation- and GB-mediated deformation.

The line broadening analysis based on the SLM revealed that, for the shear-dominated plasticity emerging in the SCS, the size-Lorentzian / strain-Gaussian separation does not work properly. This effect could only be identified by the SLM, which relies on one individual (hkl) reflection. It was shown that especially for those combinations of (hkl) plane and ϕ direction, with the strongest influence from plastic deformation, the separation fails. If the strong grain-to-grain interactions, which are characteristic for NC metals, could be modeled accurately, a model could be developed and incorporated into a modified line broadening analysis. A similar approach was pursued in order to include the strains caused by dislocations by introducing dislocation contrast factors to "modified"

Williamson-Hall and Warren-Averbach analyses [Ungar and Borbely, 1996].

To gain deeper insight in the details of the deformation behavior, two practical suggestions are proposed as follow ups to this thesis:

- The XRD transmission geometry setups used in this thesis definitely have strong advantages, e.g. continuous testing and recording of diffraction patterns since the sample is not rotated. As a result, the obtained Debye-Scherrer rings "only" yield 2D information and no complete 3D information is gathered. In order to be able to better compare the observed deformation textures to the well-established data base existent for textures in CG metals [Wassermann, 1962; Barrett and Massalski, 1966; Gambin, 2001], also in a quantitative way, the full orientation distribution function (ODF) would be necessary. This could be accomplished by additional *ex situ* XRD measurements of predeformed samples, in order to be able to rotate the sample. An alternative approach is the approximation of a full ODF, in case of materials with high crystal symmetry (such as FCC metals), from single diffraction patterns containing several Debye-Scherrer rings [Wenk and Grigull, 2003]. This would allow monitoring of the formation of the texture *in situ*, however, since assumptions must be made, the verification with additional *ex situ* experiments is mandatory. Nevertheless, the detailed texture analysis would help with the characterization of dislocation-mediated plasticity, e.g. from the exact shares of the (h00) and (hhh) texture components, the cross-slip behavior of dislocations [Brown, 1961] and consequently the relevance of full and partial dislocations could be evaluated more elaborately.

- The comparison of XRD and ACOM/TEM data should be intensified.

Therefore, ACOM/TEM analysis has to be further developed in order to be able to perform the analysis in a (hkl) selective manner, implying improved statistics. Thereby it would be possible to identify (hkl) selective processes, such as selective grain growth or selective grain shape changes.

Finally, since a comprehensive groundwork has been created on the deformation mechanisms of NC metals and alloys, the experimental methodology can also be employed to shed light on other material systems. In principle all crystalline materials with a coherent scattering domain size in the UFG or NC regime could be tested. Also the transition from the NC to the amorphous regime, mentioned at the end of the previous section, would be of great interest. Apparently, one could also think of other crystal structures (e.g. BCC, HCP) or more complex structures, such as bi-modal microstructures, nanotwinned microstructures, or multilayered composites. Furthermore, the analysis routine and the principle setup geometry could also be used for heating or irradiation experiments.

In fact, additional preliminary experiments have already been carried out on various material systems. Particularly for samples with an initial in-plane texture, it was demonstrated that the used setup, comprising an area detector in transmission geometry, allows the identification of grain rotation, grain growth, and/or detwinning from the diffraction patterns for different material systems [Funk et al., 2012; Burger, 2012]. *Ex situ* XRD, using a microfocused X-ray beam ($8 \times 20\,\mu m^2$), enables the scanning of samples which deform very locally, and thereby can help to identify different deformation modes in different spatial areas (see Appendix E for further information [Hodge et al., 2012]).

7. Summary

In NC metals, the classical deformation mechanisms of intragranular dislocation glide, multiplication, and interaction are strongly constrained due to microstructural confinement, and different deformation mechanisms may become dominant. Especially in the grain size regime below 30 nm a change from dislocation-mediated to grain boundary-mediated plasticity is expected. In this thesis, the deformation mechanisms emerging in various NC metals and alloys during mechanical testing have been investigated systematically.

Bulk samples under compressive and shear-compressive loading and thin films adherent to compliant polymer substrate under tensile loading were investigated. The bulk samples were electrodeposited Ni ($D \approx 30$ nm) and various PdAu alloys ($D \approx 10$ - 15 nm) fabricated by inert-gas condensation, whereas the substrate-supported thin films consisted of pure Pd and PdAu alloys ($D \approx 20$ - 30 nm) fabricated by sputter-deposition.

The main experimental technique was synchrotron-based XRD during mechanical testing. Dependent on the specific sample type, geometry, microstructure, and loading condition, different setups at varying beamlines were developed. Compressive and shear-compressive testing of bulk samples, as well as the improvement of thin film sample preparation allowed to achieve high plastic strains and explore the macroplastic regime. The diffraction profiles were analyzed by sophisticated peak shape analysis involving several peak parameters, including position, breadth, intensity, and asymmetry, which allowed to unravel different deformation modes. The use of area detectors in transmission geometry allowed to analyze several complete Debye-Scherrer rings, and thereby gather in-plane information of the probed microstructure. In doing so, the formation of deformation textures and the evolution of in-plane grain shape could be monitored. The knowledge of both contributed significantly to separate and untangle coexisting mechanisms; e.g. intragranular dislocation plasticity was identified by several independent features: Deformation texture formation, reversal of (hkl)-independent peak asymmetry, and the generation of an elliptic grain shape. In contrast, stress-induced GB migration was detected by isotropic grain growth. Complementary microscopic investigations employing ACOM/TEM, SEM, and optical microscopy were used to confirm the *in situ* XRD observations.

All tested samples show a succession and coexistence of several different deformation mechanisms, regardless of grain size, loading condition, or sample geometry. Large lattice strains on the order of 1% are characteristic for NC metals given by the high strength. As a result, pronounced and complex grain-to-grain interactions emerge, owing to the combination of inhomogeneous elastic lattice response and the impediment of intragranular plasticity. Consequently, accommodation processes in GBs and triple lines are necessary from the beginning, leading to increasing integral breadth and evolution of asymmetric peak shapes.

Stress-induced grain growth was observed for all samples. However, the magnitude of grain growth and its relative contribution to overall deformation strongly varied from sample type to sample type. This was attributed to differently "clean" preparation routes inducing different impurity contents. The following ranking became apparent: Coupled grain boundary migration contributes the most to overall deformation for sputter-deposited thin films. For IGC-fabricated PdAu alloys the relative contribution is clearly reduced, and for PED Ni even lower contributions were determined. By comparing the varying relative contributions of GB migration between the three material systems and among the different alloy compositions, it is found that differences in the content of light element impurities (e.g. oxygen), arising from different preparation routes, affects the relative contributions of GB migration much more pronouncedly than the pinning effect of solute atoms.

Intragranular dislocation plasticity contributed considerably to overall deformation for Ni an Pd, but was severely impeded for IGC-fabricated PdAu alloys. This observation is attributed to the very small grain size of 10 - 15 nm hampering intragranular plasticity. Comparing Ni and Pd with similar grain sizes ($D \approx 30$ nm), Pd showed reduced dislocation activity and increased GB migration. Shear-compressive loading promotes intragranular dislocation plasticity, which was inferred from stronger texture formation.

Alloying effects were identified for the continuously miscible PdAu alloy systems. For the IGC-fabricated alloys, intragranular dislocation plasticity and coupled GB motion were more pronounced for alloys with high Au contents, whereas a more brittle behavior involving crack formation was observed for low Au contents. The thin film alloys revealed a reduced grain size and an increased fiber texture compared to pure Pd. During deformation no increased resistance to grain growth was observed for the alloys. Owing to the increased fiber texture, the elastic lattice response was less inhomogeneous than for the rather isotropic pure Pd film. As a consequence of the more homogeneous microplastic deformation and the reduced initial grain size, the onset of macroplastic deformation, governed by intragranular plasticity, was deferred to higher strains.

Overall, the established experimental setups and the data analysis routines provide a powerful tool for investigating the deformation behavior of any kind of polycrystalline sample with $D \leq 1$ µm, and moreover allow a transfer to heating/cooling or irradiation experiments.

References

Ames, M., Grewer, M., Braun, C., and Birringer, R. (2012). Nanocrystalline metals go ductile under shear deformation. *Materials Science and Engineering: A*, 546(0):248 – 257.

Ames, M., Markmann, J., and Birringer, R. (2010). Mechanical testing via dominant shear deformation of small-sized specimen. *Materials Science and Engineering: A*, 528(1):526–532.

Ames, M., Markmann, J., Karos, R., Michels, A., Tschoepe, A., and Birringer, R. (2008). Unraveling the nature of room temperature grain growth in nanocrystalline materials. *Acta Materialia*, 56(16):4255–4266.

Arzt, E. (1998). Size effects in materials due to microstructural and dimensional constraints: a comparative review. *Acta Materialia*, 46(16):5611–5626.

Ashby, M. and Verrall, R. (1973). Diffusion-accommodated flow and superplasticity. *Acta Metallurgica*, 21(2):149–163.

Bachmann, F., Hielscher, R., and Schaeben, H. (2011). Grain detection from 2d and 3d ebsd data - specification of the mtex algorithm. *Ultramicroscopy*, 111(12):1720–1733.

Bachurin, D. and Gumbsch, P. (2010). Accommodation processes during deformation of nanocrystalline palladium. *Acta Materialia*, 58(16):5491–5501.

Balogh, L., Ribarik, G., and Ungar, T. (2006). Stacking faults and twin boundaries in fcc crystals determined by x-ray diffraction profile analysis. *Journal of Applied Physics*, 100(2):023512.

Barrett, C. S. and Massalski, T. B. (1966). *Structure of Metals*. McGraw-Hill, New York, ed. 3 edition.

Beck, G., editor (1995). *Edelmetall-Taschenbuch*. Huethig, Heidelberg, 2. edition.

Bei, H., Shim, S., George, E., Miller, M., Herbert, E., and Pharr, G. (2007). Compressive strengths of molybdenum alloy micro-pillars prepared using a new technique. *Scripta Materialia*, 57(5):397–400.

Bei, H., Shim, S., Pharr, G. M., and George, E. P. (2008). Effects of pre-strain on the compressive stress-strain response of mo-alloy single-crystal micropillars. *Acta Materialia*, 56(17):4762–4770.

Birringer, R. (1989). Nanocrystalline materials. *Materials Science and Engineering A*, 117(C):33–43.

Birringer, R., Gleiter, H., Klein, H. P., and Marquardt, P. (1984). Nanocrystalline materials an approach to a novel solid structure with gas-like disorder? *Physics Letters A*, 102(8):365–369.

Bitzek, E., Brandl, C., Derlet, P. M., and Van Swygenhoven, H. (2008). Dislocation Cross-Slip in nanocrystalline fcc metals. *Physical Review Letters*, 100(23):235501.

Bohm, J. (2004). *In situ tensile testing at the limits of X-ray diffraction - a new synchrotron-based technique*. PhD thesis, Universität Stuttgart.

Bohm, J., Gruber, P., Spolenak, R., Stierle, A., Wanner, A., and Arzt, E. (2004). Tensile testing of ultrathin polycrystalline films: A synchrotron-based technique. *Review of Scientific Instruments*, 75(4):1110.

Brandstetter, S., Derlet, P., Van Petegem, S., and Van Swygenhoven, H. (2008). Williamson-Hall anisotropy in nanocrystalline metals: X-ray diffraction experiments and atomistic simulations. *Acta Materialia*, 56(2):165–176.

Brandstetter, S., VanSwygenhoven, H., VanPetegem, S., Schmitt, B., Maass, R., and Derlet, P. M. (2006). From micro- to macroplasticity. *Advanced Materials*, 18(12):1545–1548.

Brenner, S. S. (1956). Tensile strength of whiskers. *Journal of Applied Physics*, 27(12):1484–1491.

Brenner, S. S. (1959). Strength of gold whiskers. *Journal of Applied Physics*, 30(2):266.

Brown, N. (1961). The dependence of wire texture in fcc metals on stacking fault energy. *Transactions of the Metallurgical Society of AIME*, 221(2):236.

Budrovic, Z., Petegem, S. V., Derlet, P. M., Schmitt, B., Swygenhoven, H. V., Schafler, E., and Zehetbauer, M. (2005). Footprints of deformation mechanisms during in situ x-ray diffraction: Nanocrystalline and ultrafine grained ni. *Applied Physics Letters*, 86(23):231910.

Budrovic, Z., Van Swygenhoven, H., Derlet, P. M., Van Petegem, S., and Schmitt, B. (2004). Plastic deformation with reversible peak broadening in nanocrystalline nickel. *Science*, 304(5668):273 –276.

Burger, S. (2012). Phd thesis. Karlsruhe Institute of Technology, unpublished.

Cahn, J. W., Mishin, Y., and Suzuki, A. (2006). Coupling grain boundary motion to shear deformation. *Acta Materialia*, 54:4953–4975.

Castrup, A. (2012). Phd thesis. Technische Universität Darmstadt, unpublished.

Castrup, A., Kuebel, C., Scherer, T., and Hahn, H. (2011). Microstructure and residual stress of magnetron sputtered nanocrystalline palladium and palladium gold films on polymer substrates. *Journal of Vacuum Science & Technology A*, 29(2).

Chen, M., Ma, E., Hemker, K. J., Sheng, H., Wang, Y., and Cheng, X. (2003). Deformation twinning in nanocrystalline aluminum. *Science*, 300(5623):1275–1277.

Cheng, S., Stoica, A. D., Wang, X., Ren, Y., Almer, J., Horton, J. A., Liu, C. T., Clausen, B., Brown, D. W., Liaw, P. K., and Zuo, L. (2009). Deformation crossover: From nano- to mesoscale. *Physical Review Letters*, 103(3):035502.

Chokshi, A., Rosen, A., Karch, J., and Gleiter, H. (1989). On the validity of the hall-petch relationship in nanocrystalline materials. *Scripta Metallurgica*, 23(10):1679–1683.

Choo, R., Toguri, J., El-Sherik, A., and Erb, U. (1995). Mass transfer and electrocrystallization analyses of nanocrystalline nickel production by pulse plating. *Journal of Applied Electrochemistry*, 25(4).

Chookajorn, T., Murdoch, H. A., and Schuh, C. A. (2012). Design of stable nanocrystalline alloys. *Science*, 337(6097):951–954.

Coble, R. (1963). A model for boundary diffusion controlled creep in polycrystalline materials. *Journal of Applied Physics*, 34(6):1679–1682.

Daniels, J. E. and Drakopoulos, M. (2009). High-energy x-ray diffraction using the pixium 4700 flat-panel detector. *Journal of Synchrotron Radiation*, 16(4):463–468.

de Keijser, T. H., Langford, J. I., Mittemeijer, E. J., and Vogels, A. B. P. (1982). Use of the voigt function in a single-line method for the analysis of x-ray diffraction line broadening. *Journal of Applied Crystallography*, 15(3):308–314.

Derlet, P., Hasnaoui, A., and Van Swygenhoven, H. (2003a). Atomistic simulations as guidance to experiments. *Scripta Materialia*, 49(7):629–635.

Derlet, P. M., Van Swygenhoven, H., and Hasnaoui, A. (2003b). Atomistic simulation of dislocation emission in nanosized grain boundaries. *Philosophical Magazine*, 83(31-34):3569–3575.

Detor, A. J. and Schuh, C. A. (2007). Tailoring and patterning the grain size of nanocrystalline alloys. *Acta Materialia*, 55(1):371–379.

Dorogoy, A. and Rittel, D. (2005). Numerical validation of the shear compression specimen. part i: Quasi-static large strain testing. *Experimental Mechanics*, 45(2):167–177.

Eberl, C. (2010). Digital image correlation and tracking. *http://www.mathworks.com/matlabcentral/fileexchange/12413-digital-image-correlation-and-tracking*.

El-Sherik, A., Erb, U., and Page, J. (1997). Microstructural evolution in pulse plated nickel electrodeposits. *Surface and Coatings Technology*, 88(1–3):70–78.

Fan, G., Li, L., Yang, B., Choo, H., Liaw, P., Saleh, T., Clausen, B., and Brown, D. (2009). In situ neutron-diffraction study of tensile deformation of a bulk nanocrystalline alloy. *Materials Science and Engineering: A*, 506(1–2):187–190.

Fan, G. J., Fu, L. F., Wang, Y. D., Ren, Y., Choo, H., Liaw, P. K., Wang, G. Y., and Browning, N. D. (2006a). Uniaxial tensile plastic deformation of a bulk nanocrystalline alloy studied by a high-energy x-ray diffraction technique. *Applied Physics Letters*, 89(10):101918.

Fan, G. J., Wang, Y. D., Fu, L. F., Choo, H., Liaw, P. K., Ren, Y., and Browning, N. D. (2006b). Orientation-dependent grain growth in a bulk nanocrystalline alloy during the uniaxial compressive deformation. *Applied Physics Letters*, 88(17):171914.

Funk, M., Kennerknecht, T., Lohmiller, J., Gruber, P. A., Kobler, A., Zhang, X., and Eberl, C. (2012). to be submitted to Acta Materialia.

Gambin, W. (2001). *Plasticity and textures*. Kluwer Academic, Dordrecht. Includes bibliographical references and index.

Geim, A. K. and Novoselov, K. S. (2007). The rise of graphene. *Nature Materials*, 6(3):183–191.

Genzel, C. (1997). X-Ray stress gradient analysis in thin layers - problems and attempts at their solution. *physica status solidi (a)*, 159(2):283–296.

Gianola, D., Mendis, B., Cheng, X., and Hemker, K. (2008). Grain-size stabilization by impurities and effect on stress-coupled grain growth in nanocrystalline al thin films. *Materials Science and Engineering: A*, 483–484(0):637–640.

Gianola, D., Van Petegem, S., Legros, M., Brandstetter, S., Van Swygenhoven, H., and Hemker, K. (2006). Stress-assisted discontinuous grain growth and its effect on the deformation behavior of nanocrystalline aluminum thin films. *Acta Materialia*, 54(8):2253–2263.

Gianola, D. S., Sedlmayr, A., Moenig, R., Volkert, C. A., Major, R. C., Cyrankowski, E., Asif, S. A. S., Warren, O. L., and Kraft, O. (2011). In situ nanomechanical testing in focused ion beam and scanning electron microscopes. *Review of Scientific Instruments*, 82(6):063901.

Gleiter, H. (1984). Nanocrystalline structures - an approach to new materials. *Zeitschrift für Metallkunde*, 75(4):263–267.

Grewer, M., Markmann, J., Karos, R., Arnold, W., and Birringer, R. (2011). Shear softening of grain boundaries in nanocrystalline pd. *Acta Materialia*, 59(4):1523–1529.

Gross, D., Hauger, W., and Schnell, W. (2005). *Technische Mechanik*, volume 2: Elastostatik. Springer, Berlin, 8., erw. aufl. edition.

Gruber, P. A., Böhm, J., Onuseit, F., Wanner, A., Spolenak, R., and Arzt, E. (2008a). Size effects on yield strength and strain hardening for ultra-thin cu films with and without passivation: A study by synchrotron and bulge test techniques. *Acta Materialia*, 56(10):2318–2335.

Gruber, P. A., Olliges, S., Arzt, E., and Spolenak, R. (2008b). Temperature dependence of mechanical properties in ultrathin au films with and without passivation. *Journal of Materials Research*, 23(9):2406–2419.

Gruber, P. A., Solenthaler, C., Arzt, E., and Spolenak, R. (2008c). Strong single-crystalline au films tested by a new synchrotron technique. *Acta Materialia*, 56(8):1876–1889.

Hadian, S. and Gabe, D. (1999). Residual stresses in electrodeposits of nickel and nickel-iron alloys. *Surface and Coatings Technology*, 122(2–3):118–135.

Hahn, H., Mondal, P., and Padmanabhan, K. (1997). Plastic deformation of nanocrystalline materials. *Nanostructured Materials*, 9(1-8):603–606.

Hall, E. O. (1951). The deformation and ageing of mild steel: Iii discussion of results. *Proceedings of the Physical Society. Section B*, 64(9):747.

Herring, C. (1950). Diffusional viscosity of a polycrystalline solid. *Journal of Applied Physics*, 21(5):437–445.

Hodge, A., Furnish, T., Navid, A., and Barbee Jr., T. (2011). Shear band formation and ductility in nanotwinned cu. *Scripta Materialia*, 65(11):1006–1009.

Hodge, A., Wang, Y., and Barbee Jr., T. (2006). Large-scale production of nano-twinned, ultrafine-grained copper. *Materials Science and Engineering: A*, 429(1–2):272–276.

Hodge, A., Wang, Y., and Barbee Jr., T. (2008). Mechanical deformation of high-purity sputter-deposited nano-twinned copper. *Scripta Materialia*, 59(2):163–166.

Hodge, A. M., Lohmiller, J., Gruber, P. A., and Furnish, T. A. (2012). to be submitted to Applied Physics Letters.

Idrissi, H., Wang, B., Colla, M. S., Raskin, J. P., Schryvers, D., and Pardoen, T. (2011). Ultrahigh strain hardening in thin palladium films with nanoscale twins. *Advanced Materials*, 23(18):2119–2122.

Ivanisenko, Y., Kurmanaeva, L., Weissmueller, J., Yang, K., Markmann, J., Roesner, H., Scherer, T., and Fecht, H. (2009). Deformation mechanisms in nanocrystalline palladium at large strains. *Acta Materialia*, 57(11):3391–3401.

Jin, Z., Gumbsch, P., Albe, K., Ma, E., Lu, K., Gleiter, H., and Hahn, H. (2008). Interactions between non-screw lattice dislocations and coherent twin boundaries in face-centered cubic metals. *Acta Materialia*, 56(5):1126–1135.

Jin, Z., Gumbsch, P., Ma, E., Albe, K., Lu, K., Hahn, H., and Gleiter, H. (2006). The interaction mechanism of screw dislocations with coherent twin boundaries in different face-centred cubic metals. *Scripta Materialia*, 54(6):1163–1168.

Johanns, K. E., Sedlmayr, A., Sudharshan Phani, P., Moenig, R., Kraft, O., George, E. P., and Pharr, G. M. (2012). In-situ tensile testing of single-crystal molybdenum-alloy fibers with various dislocation densities in a scanning electron microscope. *Journal of Materials Research*, 27(03):508–520.

Kaischew, R. (1967). Keimbildung bei der elektrolytischen metallabscheidung. *Chemie Ingenieur Technik*, 39(9–10):554–562.

Keller, R., Baker, S. P., and Arzt, E. (1998). Quantitative analysis of strengthening mechanisms in thin cu films: Effects of film thickness, grain size, and passivation. *Journal of Materials Research*, 13(05):1307–1317.

Klug, Harold P. ; Alexander, L. E. (1974). *X-ray diffraction procedures: for polycrystalline and amorphous materials*. A Wiley-Interscience publication. Wiley, New York [u.a.], 2. ed. edition.

Koch, C. C., Scattergood, R. O., Darling, K. A., and Semones, J. E. (2008). Stabilization of nanocrystalline grain sizes by solute additions. *Journal of Materials Science*, 43(23-24):7264–7272.

Kraft, O., Gruber, P. A., Moenig, R., and Weygand, D. (2010). Plasticity in confined dimensions. In Clarke, D. R., Ruhle, M., and Zok, F., editors, *Annual Review of Materials Research, Vol 40*, volume 40, pages 293–317. Annual Reviews, Palo Alto.

Kumar, K. S., Van Swygenhoven, H., and Suresh, S. (2003). Mechanical behavior of nanocrystalline metals and alloys. *Acta Materialia*, 51(19):5743–5774.

Kurmanaeva, L., Ivanisenko, Y., Markmann, J., Yang, K., Fecht, H.-J., and Weissmueller, J. (2010). Work hardening and inherent plastic instability of nanocrystalline metals. *physica status solidi (RRL) - Rapid Research Letters*, 4(5-6):130–132.

Labusch, R. (1970). A statistical theory of solid solution hardening. *physica status solidi (b)*, 41(2):659–669.

Langford, J. I. (1978). A rapid method for analysing the breadths of diffraction and spectral lines using the voigt function. *Journal of Applied Crystallography*, 11(1):10–14.

Legros, M., Gianola, D. S., and Hemker, K. J. (2008). In situ TEM observations of fast grain-boundary motion in stressed nanocrystalline aluminum films. *Acta Materialia*, 56(14):3380–3393.

Li, H., Choo, H., Ren, Y., Saleh, T. A., Lienert, U., Liaw, P. K., and Ebrahimi, F. (2008). Strain-Dependent deformation behavior in nanocrystalline metals. *Physical Review Letters*, 101(1):015502.

Li, T., Huang, Z., Xi, Z., Lacour, S., Wagner, S., and Suo, Z. (2005). Delocalizing strain in a thin metal film on a polymer substrate. *Mechanics of Materials*, 37(2-3):261–273.

Lohmiller, J. (2009). High ductility in thin metal films for flexible electronics. Diploma thesis, Universität Karlsruhe, TH, Germany.

Lohmiller, J., Baumbusch, R., Kerber, M. B., Castrup, A., Hahn, H., Schafler, E., Zehetbauer, M., Kraft, O., and Gruber, P. A. (2012a). Following the deformation behavior of nanocrystalline pd films on polyimide substrates using in situ synchrotron xrd. accepted by Mechanics of Materials.

Lohmiller, J., Baumbusch, R., Kraft, O., and Gruber, P. A. (2013). Differentiation of deformation modes in nanocrystalline pd films inferred from peak asymmetry evolution using in situ x-ray diffraction. *Physical Review Letters*, 110(6):066101.

Lohmiller, J., Grewer, M., Braun, C., Kobler, A., Kuebel, C., Schueler, K., Honkimäki, V., Hahn, H., Kraft, O., Birringer, R., and Gruber, P. A. (2012b). Untangling dislocation and grain boundary mediated plasticity in nanocrystalline nickel. submitted to Acta Materialia.

Lohmiller, J., Woo, N. C., and Spolenak, R. (2010). Microstructure-property relationship in highly ductile Au-Cu thin films for flexible electronics. *Materials Science and Engineering: A*, 527(29-30):7731–7740.

Lu, L., Chen, X., Huang, X., and Lu, K. (2009). Revealing the maximum strength in nanotwinned copper. *Science*, 323(5914):607–610.

Lu, N., Wang, X., Suo, Z., and Vlassak, J. (2007). Metal films on polymer substrates stretched beyond 50%. *Applied Physics Letters*, 91(22):221909–221909–3.

Ma, E. (2004). Watching the nanograins roll. *Science*, 305(5684):623 –624.

Ma, E., Wang, Y. M., Lu, Q. H., Sui, M. L., Lu, L., and Lu, K. (2004). Strain hardening and large tensile elongation in ultrahigh-strength nano-twinned copper. *Applied Physics Letters*, 85:4932.

Markmann, J., Bachurin, D., Shao, L., Gumbsch, P., and Weissmueller, J. (2010). Microstrain in nanocrystalline solids under load by virtual diffraction. *EPL (Europhysics Letters)*, 89(6):66002.

Markmann, J., Bunzel, P., Roesner, H., Liu, K. W., Padmanabhan, K. A., Birringer, R., Gleiter, H., and Weissmueller, J. (2003). Microstructure evolution during rolling of inert-gas condensed palladium. *Scripta Materialia*, 49(7):637–644.

McFadden, S. X., Mishra, R. S., Valiev, R. Z., Zhilyaev, A. P., and Mukherjee, A. K. (1999). Low-temperature superplasticity in nanostructured nickel and metal alloys. *Nature*, 398(6729):684–686.

Meyers, M., Mishra, A., and Benson, D. (2006). Mechanical properties of nanocrystalline materials. *Progress in Materials Science*, 51:427–556.

Millett, P. C., Selvam, R. P., and Saxena, A. (2007). Stabilizing nanocrystalline materials with dopants. *Acta Materialia*, 55(7):2329–2336.

Mittemeijer, E. J. and Welzel, U. (2008). The state of the art of the diffraction analysis of crystallitesize and lattice strain. *Zeitschrift für Kristallographie*, 223(9):552–560.

Moti, E., Shariat, M. H., and Bahrololoom, M. E. (2008). Influence of cathodic overpotential on grain size in nanocrystalline nickel deposition on rotating cylinder electrodes. *Journal of Applied Electrochemistry*, 38(5):605–612. WOS:000254804700004.

Motz, C., Weygand, D., Senger, J., and Gumbsch, P. (2009). Initial dislocation structures in 3-D discrete dislocation dynamics and their influence on microscale plasticity. *Acta Materialia*, 57(6):1744–1754.

Natter, H., Schmelzer, M., and Hempelmann, R. (1998). Nanocrystalline nickel and nickel-copper alloys: Synthesis, characterization, and thermal stability. *Journal of Materials Research*, 13(05):1186–1197.

Olliges, S., Gruber, P., Auzelyte, V., Ekinci, Y., Solak, H., and Spolenak, R. (2007). Tensile strength of gold nanointerconnects without the influence of strain gradients. *Acta Materialia*, 55(15):5201–5210.

Petch, N. (1953). The cleavage strength of polycrystals. *Journal Iron Steel Institute*, 58(174):25–28.

Richter, G., Hillerich, K., Gianola, D. S., Moenig, R., Kraft, O., and Volkert, C. A. (2009). Ultrahigh strength single crystalline nanowhiskers grown by physical vapor deposition. *Nano Lett.*, 9(8):3048–3052.

Roesner, H., Markmann, J., and Weissmueller, J. (2004). Deformation twinning in nanocrystalline pd. *Philosophical Magazine Letters*, 84:321–334.

Rupert, T. J., Gianola, D. S., Gan, Y., and Hemker, K. J. (2009). Experimental observations of Stress-Driven grain boundary migration. *Science*, 326(5960):1686–1690.

Saada, G. (2005). Hall-Petch revisited. *Materials Science and Engineering: A*, 400–401(0):146–149.

Saada, G. and Kruml, T. (2011). Deformation mechanisms of nanograined metallic polycrystals. *Acta Materialia*, 59:2565–2574.

Scattergood, R., Koch, C., Murty, K., and Brenner, D. (2008). Strengthening mechanisms in nanocrystalline alloys. *Materials Science and Engineering: A*, 493(1–2):3–11.

Schaefer, J., Stukowski, A., and Albe, K. (2011). Plastic deformation of nanocrystalline PdAu alloys: On the interplay of grain boundary solute segregation, fault energies and grain size. *Acta Materialia*, 59:2957–2968.

Schiotz, J. and Jacobsen, K. W. (2003). A maximum in the strength of nanocrystalline copper. *Science*, 301(5638):1357–1359.

Schmitt, B., Broennimann, C., Eikenberry, E. F., Gozzo, F., Hoermann, C., Horisberger, R., and Patterson, B. (2003). Mythen detector system. *Nuclear Instruments and Methods in Physics Research A*, 501:267–272.

Sedlmayr, A., Bitzek, E., Gianola, D. S., Richter, G., Mönig, R., and Kraft, O. (2012). Existence of two twinning-mediated plastic deformation modes in au nanowhiskers. *Acta Materialia*, 60(9):3985–3993.

Senger, J., Weygand, D., Gumbsch, P., and Kraft, O. (2008). Discrete dislocation simulations of the plasticity of micro-pillars under uniaxial loading. *Scripta Materialia*, 58(7):587–590.

Senger, J., Weygand, D., Motz, C., Gumbsch, P., and Kraft, O. (2011). Aspect ratio and stochastic effects in the plasticity of uniformly loaded micrometer-sized specimens. *Acta Materialia*, 59(8):2937–2947.

Shan, Z., Stach, E. A., Wiezorek, J. M. K., Knapp, J. A., Follstaedt, D. M., and Mao, S. X. (2004). Grain Boundary-Mediated plasticity in nanocrystalline nickel. *Science*, 305(5684):654 –657.

Shen, Y., Lu, L., Lu, Q., Jin, Z., and Lu, K. (2005). Tensile properties of copper with nano-scale twins. *Scripta Materialia*, 52(10):989 – 994.

Singh, A. K. and Balasingh, C. (2001). X-ray diffraction line broadening under elastic deformation of a polycrystalline sample: An elastic-anisotropy effect. *Journal of Applied Physics*, 90(5):2296–2302.

Slater, J. C. (1964). Atomic radii in crystals. *The Journal of Chemical Physics*, 41(10):3199–3204.

Tang, F., Gianola, D., Moody, M., Hemker, K., and Cairney, J. (2012). Observations of grain boundary impurities in nanocrystalline al and their influence on microstructural stability and mechanical behaviour. *Acta Materialia*, 60(3):1038–1047.

Thompson, C. (1993). The yield stress of polycrystalline thin films. *Journal of Materials Research*, 8(02):237–238.

Toraya, H. (1990). Array-type universal profile function for powder pattern fitting. *Journal of Applied Crystallography*, 23:485–491.

Trelewicz, J. R. and Schuh, C. A. (2007). The Hall-Petch breakdown in nanocrystalline metals: A crossover to glass-like deformation. *Acta Materialia*, 55(17):5948–5958.

Uchic, M. D., Dimiduk, D. M., Florando, J. N., and Nix, W. D. (2004). Sample dimensions influence strength and crystal plasticity. *Science*, 305(5686):986–989.

Underwood, E. E. (1970). *Quantitative Stereology*. Addison-Wesley Publishing Co., Reading, MA.

Ungar, T. and Borbely, A. (1996). The effect of dislocation contrast on x-ray line broadening: A new approach to line profile analysis. *Applied Physics Letters*, 69(21):3173.

Ungar, T., Schafler, E., Hanak, P., Bernstorff, S., and Zehetbauer, M. (2005). Vacancy concentrations determined from the diffuse background scattering of x-rays in plastically deformed copper. *Zeitschrift fuer Metallkunde/Materials Research and Advanced Techniques*, 96(6):578–583.

Ungar, T., Schafler, E., Hanak, P., Bernstorff, S., and Zehetbauer, M. (2007). Vacancy production during plastic deformation in copper determined by in situ x-ray diffraction. *Materials Science and Engineering: A*, 462(1–2):398–401.

Van Swygenhoven, H. and Derlet, P. (2001). Grain-boundary sliding in nanocrystalline fcc metals. *Physical Review B*, 64(22).

Van Swygenhoven, H., Derlet, P. M., and Froseth, A. G. (2004). Stacking fault energies and slip in nanocrystalline metals. *Nat Mater*, 3(6):399–403.

Van Swygenhoven, H., Schmitt, B., Derlet, P. M., Van Petegem, S., Cervellino, A., Budrovic, Z., Brandstetter, S., Bollhalder, A., and Schild, M. (2006). Following peak profiles during elastic and plastic deformation: A synchrotron-based technique. *Review of Scientific Instruments*, 77(1):013902.

Van Swygenhoven, H. and Weertman, J. (2006). Deformation in nanocrystalline metals. *Materials Today*, 9(5):24–31.

Vegard, L. (1921). Die Konstitution der Mischkristalle und die Raumfüllung der Atome. *Zeitschrift für Physik A: Hadrons and Nuclei*, 5(1):17–26.

Venkatraman, R. and Bravman, J. C. (1992). Separation of film thickness and grain boundary strengthening effects in al thin films on si. *Journal of Materials Research*, 7(08):2040–2048.

Vo, N., Averback, R., Bellon, P., Odunuga, S., and Caro, A. (2008). Quantitative description of plastic deformation in nanocrystalline cu: Dislocation glide versus grain boundary sliding. *Physical Review B*, 77.

von Blanckenhagen, B., Gumbsch, P., and Arzt, E. (2003). Dislocation sources and the flow stress of polycrystalline thin metal films. *Philosophical Magazine Letters*, 83(1):1–8.

Wang, Y., Jankowski, A., and Hamza, A. (2007). Strength and thermal stability of nanocrystalline gold alloys. *Scripta Materialia*, 57(4):301 – 304.

Wang, Y. B., Li, B. Q., Sui, M. L., and Mao, S. X. (2008). Deformation-induced grain rotation and growth in nanocrystalline ni. *Applied Physics Letters*, 92(1):011903.

Wang, Y. M., Ott, R. T., Hamza, A. V., Besser, M. F., Almer, J., and Kramer, M. J. (2010). Achieving large uniform tensile ductility in nanocrystalline metals. *Physical Review Letters*, 105(21):215502.

Warren, B. (1959). X-ray studies of deformed metals. *Progress in Metal Physics*, 8:147–202.

Warren, B. E. and Averbach, B. L. (1950). The effect of cold-work distortion on x-ray patterns. *Journal of Applied Physics*, 21(6):595–599.

Wassermann, Guenter ; Grewen, J. (1962). *Texturen metallischer Werkstoffe*. Springer, Berlin, 2., neubearb. und erw. aufl. edition.

Weertman, J. R. (2012). Retaining the nano in nanocrystalline alloys. *Science*, 337(6097):921–922.

Weidner, D. J., Wang, Y., and Vaughan, M. T. (1994). Strength of diamond. *Science*, 266(5184):419–422.

Weissmueller, J., Markmann, J., Grewer, M., and Birringer, R. (2011). Kinematics of polycrystal deformation by grain boundary sliding. *Acta Materialia*, 59(11):4366–4377.

Wenk, H.-R. and Grigull, S. (2003). Synchrotron texture analysis with area detectors. *Journal of Applied Crystallography*, 36(4):1040–1049.

Weygand, D., Friedman, L. H., Giessen, E. V. d., and Needleman, A. (2002). Aspects of boundary-value problem solutions with three-dimensional dislocation dynamics. *Modelling and Simulation in Materials Science and Engineering*, 10(4):437–468.

Weygand, D. and Gumbsch, P. (2005). Study of dislocation reactions and rearrangements under different loading conditions. *Materials Science and Engineering: A*, 400–401(0):158–161.

Williamson, G. and Hall, W. (1953). X-ray line broadening from filed aluminium and wolfram. *Acta Metallurgica*, 1(1):22–31.

Xia, Z. C. and Hutchinson, J. W. (2000). Crack patterns in thin films. *Journal of the Mechanics and Physics of Solids*, 48(6-7):1107–1131.

Xiang, Y., Li, T., Suo, Z., and Vlassak, J. J. (2005). High ductility of a metal film adherent on a polymer substrate. *Applied Physics Letters*, 87(16):161910–3.

Yamakov, V., Wolf, D., Phillpot, S. R., Mukherjee, A. K., and Gleiter, H. (2002). Dislocation processes in the deformation of nanocrystalline aluminium by molecular-dynamics simulation. *Nature Materials*, 1(1):45–49.

Yamakov, V., Wolf, D., Salazar, M., Phillpot, S., and Gleiter, H. (2001). Length-scale effects in the nucleation of extended dislocations in nanocrystalline al by molecular-dynamics simulation. *Acta Materialia*, 49(14):2713 – 2722.

Zacharias, J. (1933). The temperature dependence of young's modulus for nickel. *Physical Review*, 44(2):116–122.

Zhu, Y., Liao, X., and Wu, X. (2012). Deformation twinning in nanocrystalline materials. *Progress in Materials Science*, 57(1):1–62.

A. Post Mortem ACOM Measurements

ACOM/TEM sample preparation and corresponding experiments were carried out by Aaron Kobler (INT, KIT), and detailed analysis was developed and performed together in close collaboration. In a first step TEM-lamellae were taken out of deformed and undeformed samples using a FEI Strata 400S DualBeam FIB. After final polishing of the TEM lamellae with a $5\,\mathrm{kV}$ Ga ion beam, the lamellae were investigated using a FEI Tecnai F20 SuperTwin TEM in μP-STEM mode at $200\,\mathrm{kV}$ with a spot size of 1 - 1.5 nm and a semi convergence angle of 1.4 mrad. The μP-STEM imaging was combined with the Automated Crystal Orientation and phase Mapping (ACOM) solution by Nanomegas (ASTAR).

Data pre-processing for quantification of the ACOM maps consisted of the following steps:

1. Exporting the raw data as angle files after matching the diffraction pattern with the Nanomegas software package (ASTAR).

2. Filtering of the ACOM maps with a median filter of the combined Euler angles.

3. Grain recognition in the ACOM maps was performed using MTEX [Bachmann et al., 2011] in MatLab.

4. Grain filtering was done removing grains with an area smaller than 15 pixels corresponding to an equivalent diameter $< 5.8\,\mathrm{nm}$ and grains with more than 50% of the pixels below a combined limit of 15% reliability and index (confidence) of 20 were removed.

5. The polygon representation of the remaining grains was smoothed by averaging over an angular range of $\pm 20°$ before approximating the grains by an elliptical fit.

The resulting grain representation was used to determine orientation-dependent grain sizes, yielding the in-plane grain shape, see e.g. Fig. 4.4(d). This was computed by averaging the radii of all individual grain ellipses for a specific ϕ direction and subsequent iteration for ϕ increments of $2°$. To reveal the crystal orientation density with respect to ϕ, the orientation density was determined from the filtered ACOM data (step 2) in a specific direction ϕ by integrating the pixels for a certain (hkl) crystal orientation. This was repeated for all ϕ angles in steps of $2°$ (e.g. see Fig. 4.4(e) for the (111) orientation).

B. FEM Simulations of Shear-Compression Specimen

The FEM simulations were carried out by Manuel Grewer (Universität des Saarlandes). Unlike normal tensile or compression tests, the deformation state in the gauge section of a SCS is a complex three-dimensional stress and strain state and therefore stress-strain curves are not directly accessible from the measured load-displacement data. However, it was shown in Ref. [Dorogoy and Rittel, 2005], that the equivalent stress-strain curve can be obtained using the finite element method (FEM). Essentially, a virtual representation of the in reality tested SCS is deformed using FEM in conjunction with a suitable material law. By varying the parameters of the material law, the simulation output in the load-displacement domain is refined until a best fit to the measured load-displacement data has been achieved. The so refined material parameters of the constitutive law are then used to compute the desired stress-strain relationship. Here, the FEM analysis was performed with Abaqus following the procedures described in [Ames et al., 2010]. However, instead of an exponential strain hardening term an equation proposed in Ref. [Saada and Kruml, 2011] is used, which reads:

$$\sigma(\varepsilon_p) = \sigma_m + \frac{\sigma_s(exp(\frac{\varepsilon_p}{\varepsilon_s}) - 1)}{exp(\frac{\varepsilon_p}{\varepsilon_s}) - 2q + 1} \tag{B.1}$$

and gives the better fit to the experiment. The stress σ depends on the accumulated plastic strain ε_p, σ_m is the onset of deviation from linear elasticity that is prescribed in the elastic regime, σ_s is the maximum or saturation stress, $0 < q < 1$ is a parameter that modifies the curvature of strain hardening and

$$\varepsilon_s = \frac{\sigma_s}{2E(1-q)} \tag{B.2}$$

where E is the Young's modulus.

C. Supplementary Data for Pd Thin Film Tensile Testing

In addition to section 5.2.1, more results on the deformation behavior of pure Pd thin films on compliant substrate are presented, in order to corroborate the given interpretation. It is shown in Figs. C.1 (peak position), C.2 (lattice strain), C.3 (IBR), and C.4 (asymmetry) that the general evolution of peak parameters, described in the main results of section 5.2.1.1 by means of the (111) reflection, is qualitatively equivalent for other (hkl) reflections. However, quantitative differences arise, e.g. the (200) peak shift, yield a higher lattice strain compared to (111) given by the lower stiffness of (200) planes which consequently bear more elastic strain than the stiffer (111) planes.

The results from the SLM for individual (hkl) reflections are shown in Fig. C.5. Although the absolute values vary between the different (hkl) planes, the overall trend is similar. The averaged values are displayed in Fig. 5.2 of this thesis. Comparison to grain sizes from ACOM/TEM analysis evidence that averaging of the SLM results yields reasonable values.

Fig C.6 shows the (hkl)-dependent evolution of IBR, normalized by its initial value for cycle 1. The (hhh) and (h00) reflections exhibit lower relative increase in IBR compared to other (hkl) reflections, which is in line with elastic anisotropy dominated behavior [Singh and Balasingh, 2001].

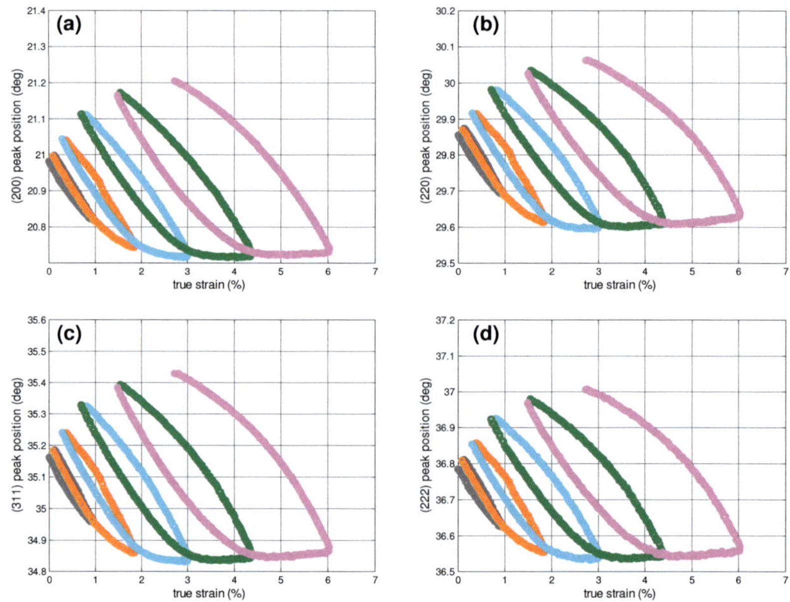

Figure C.1.: (hkl)-dependent evolution of peak position over ε. (a) (200), (b) (220), (c) (311), and (d) (222)

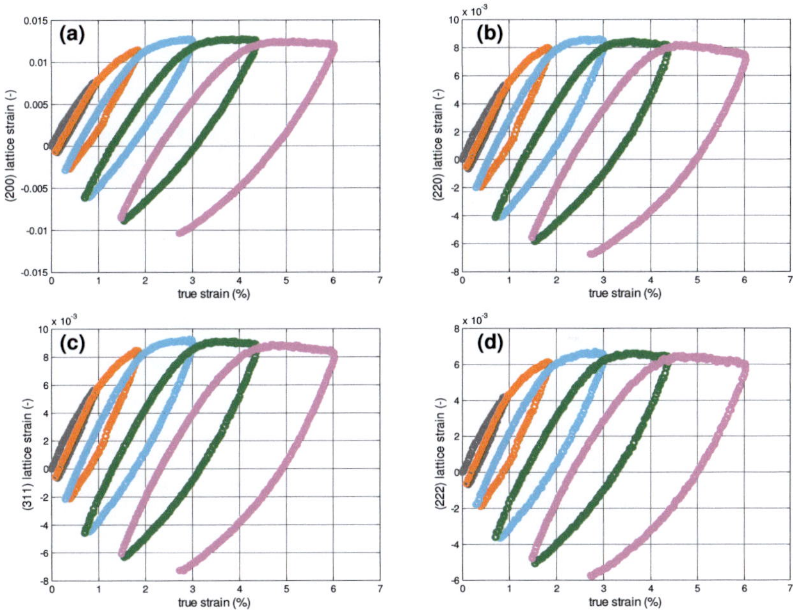

Figure C.2.: (hkl)-dependent evolution of computed lattice strain over ε. (a) (200), (b) (220), (c) (311), and (d) (222)

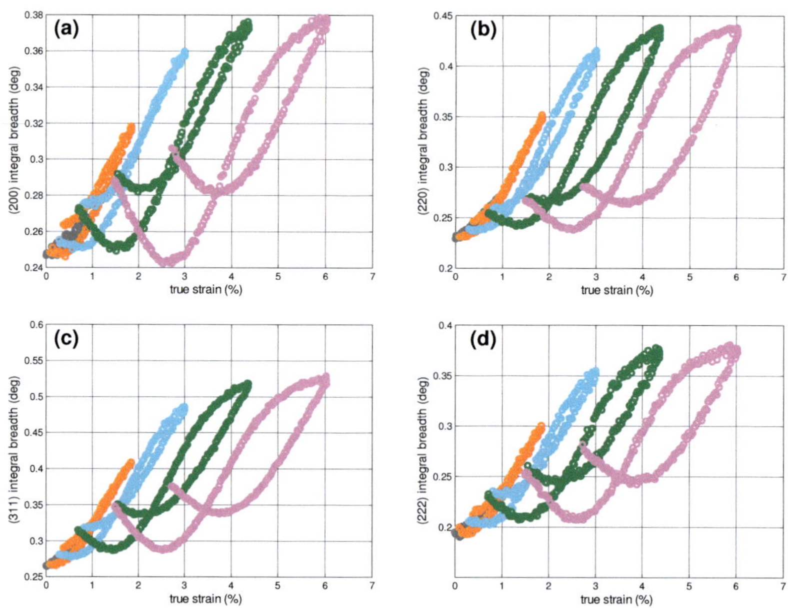

Figure C.3.: (hkl)-dependent evolution of integral peak breadth over ε. (a) (200), (b) (220), (c) (311), and (d) (222)

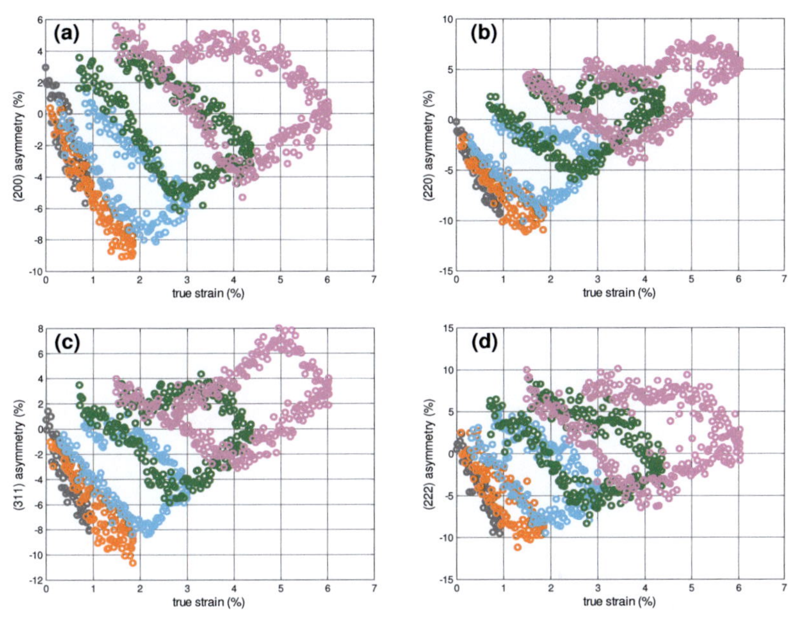

Figure C.4.: (hkl)-dependent evolution of peak asymmetry over ε. (a) (200), (b) (220), (c) (311), and (d) (222)

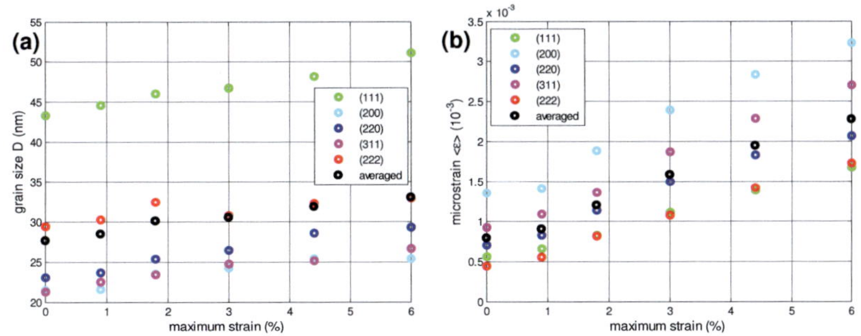

Figure C.5.: Grain size D and microstrain $<\varepsilon>$ calculated by the SLM from peak broadening data in the unloaded states. The individual reflections show similar trends, only the absolute values differ.

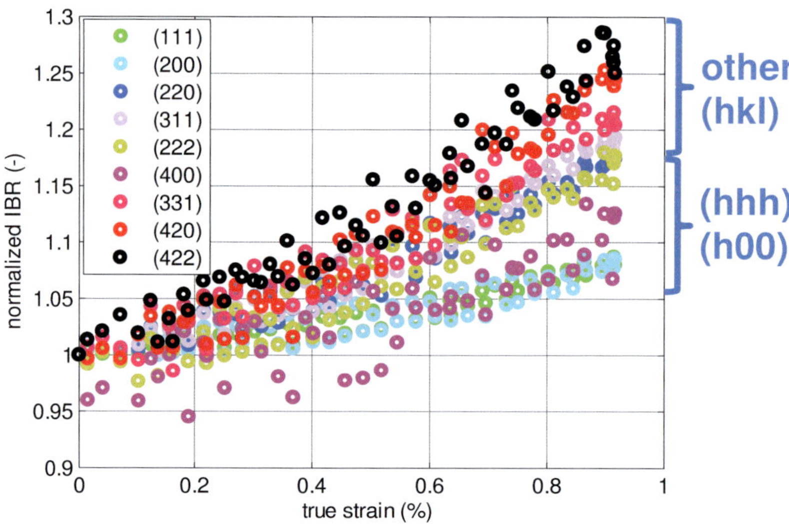

Figure C.6.: Evolution of normalized IBR for all measured reflections over ε in cycle 1.

D. AuCu Thin Film Tensile Testing

The likewise continuously miscible AuCu alloy system is an alternative to the Pd and PdAu alloys, and can serve for comparison. The NC AuCu thin films were fabricated by direct current (DC) magnetron co-sputtering. During sputter-deposition of the thin films, a composition gradient was created along the sample length by not rotating the substrate holder. For further details on fabrication, microstructure and the deformation behavior investigated by *in situ* SEM tensile testing, it is referred to Refs. [Lohmiller, 2009; Lohmiller et al., 2010]. However, it is pointed out, that the initial $<111>$ fiber texture of the AuCu alloys is much more pronounced than in Pd and PdAu alloys. The initial amplitude ratio of the (111) to (200) reflections is in the range of 12 compared to ≈ 4, which was found for Pd and PdAu alloys (cf. Table 5.1). For *in situ* tensile testing, the setups introduced in sections 3.2.2 and 3.2.3 were used with an X-ray energy of 7.97 keV. In contrast to all XRD data from the main part of this thesis, the AuCu data was fitted with an extended split-type Pearson VII function, which comprises two shape parameters, one for each side of the peak. As a result, the asymmetry parameter does not reflect the ratio of the areas of both peak halves. In order to obtain an asymmetry parameter equivalent to the asymmetry parameter used so far, the asymmetry for AuCu data is calculated according to $A = 2\theta_{POS} - 2\theta_{COM}$, which is the subtraction of the peak centroid ($2\theta_{COM}$) from the position of the peak maximum ($2\theta_{POS}$).

First, it is reported from experiments carried out at the SLS with the setup from section 3.2.2. The generalized behavior during *in situ* XRD tensile testing is illustrated in Fig. D.1. Based on the evolution of the (111) peak parameters in loading direction ($\phi = 90°$), three deformation modes are identified: (i) elastic, (ii) shear banding, and (iii) cracking. In general, the AuCu alloys are much more ductile compared to the Pd and PdAu alloys. Film cracking is only observed at high strains, high Cu contents, and/or in the annealed state [Lohmiller et al., 2010].

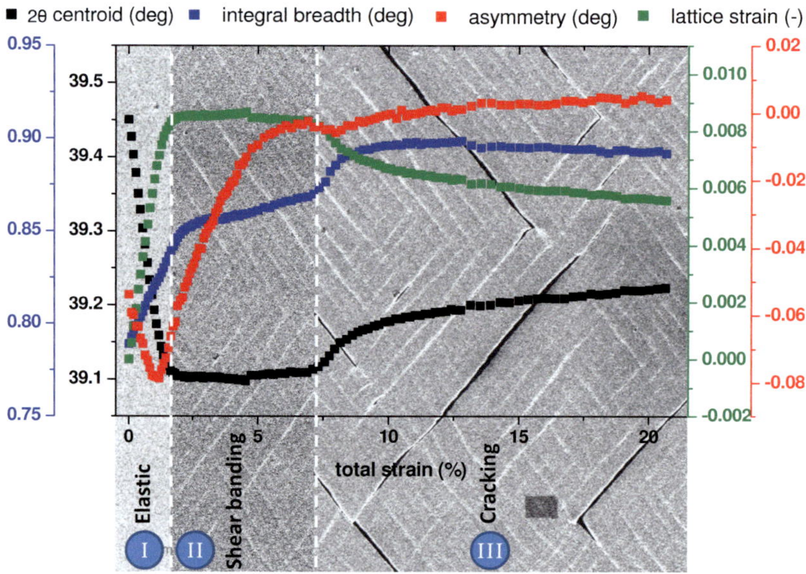

Figure D.1.: Evolution of (111) peak parameters in loading direction ($\phi = 90°$) of an annealed AuCu thin film during tensile testing. Three different deformation modes can be identified correlating with results from *in situ* SEM tensile testing.

Making use of the composition gradient, four different alloy compositions (i.e. four positions) are investigated within a single experiment. Owing to the investigation of four different positions along the sample, the straining is exceptionally applied stepwise, not continuously as during all other experiments. The strongly alloy-dependent peak parameter evolution is shown in Fig. D.2. The different initial 2θ values arise from the alloy-dependent lattice constants, while the different initial IBRs result from differences in grain size and microstrain. By applying the SLM, grain size generally decreases, while microstrain increases, for increasing Cu contents in the investigated composition range. Also the initial asymmetry values show differences, which could originate from different residual stress states. The evolution of each parameter shows an alloy-dependent behavior, and furthermore, the transition strains of the different deformation modes are clearly dependent on the alloy composition: The elastic regime slightly extends to higher strains for higher Cu contents. On the other hand, the ductile shear banding regime is strongly broadened to higher strains for lower Cu contents. As a consequence, film cracking is deferred to higher strains for alloys with higher Au contents.

In order to investigate in-plane effects in the diffraction data, samples were also tested with the setup at ANKA described in section 3.2.3, using an area detector. Since integration and read out time of the MAR-CCD are relatively high, only one spot on the composition graded samples is probed and therefore, tests are run continuously. This results in a strain rate of $\dot{\varepsilon}$ $= 2 \times 10^{-5}\,\mathrm{s}^{-1}$, which is in the same range than the stepwise tested samples, when the duration of XRD measurements is included.

In the following, results from tests on pure Au and an AuCu alloy with 10 at% Cu are compared. In Fig. D.3, the (111) lattice strain and the (111) peak asymmetry of both samples evaluated in loading direction ($\phi = 90°$) are displayed. The lattice strain of the pure sample is considerably lower. The maximum lattice strain value is even less than half of the maximum value of the alloy. However, less strain release at high plastic strains is observed in the pure sample. On the other hand, for the alloy, an asymmetry reversal can be explicitly identified, as it was also observed for pure Pd (cf. section 5.2.1.1 and Fig. 5.15 therein), while the reversal for pure Au is much less distinct.

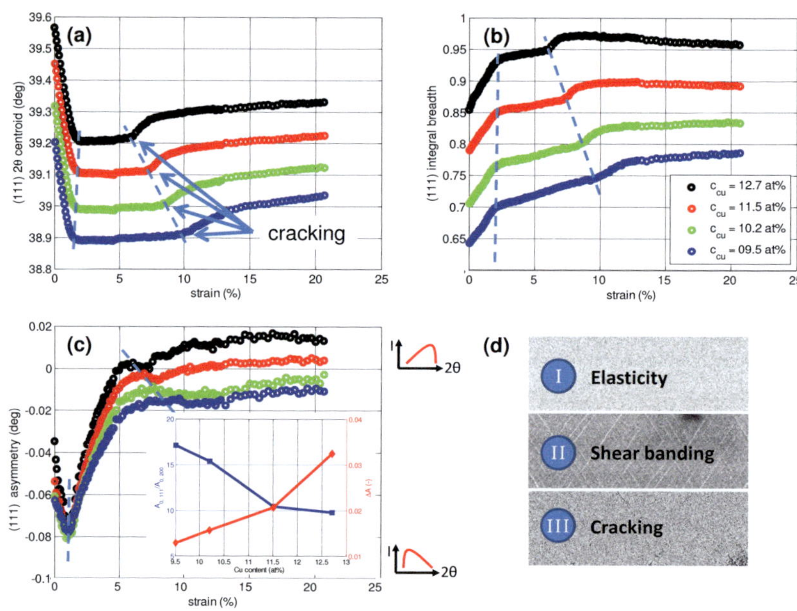

Figure D.2.: Alloy-dependent evolution of (111) peak parameters ($\phi = 90°$) of an annealed composition-graded AuCu thin film during tensile testing. Four positions / alloy compositions are probed along the samples gauge length. (a) 2θ centroid, (b) IBR, and (c) asymmetry as a function of ε. The inset in (c) demonstrates that an increasing initial <111> fiber texture correlates with a decreasing change in peak asymmetry (cf. Fig. 5.17 in section 5.3.2). The dashed lines separate the different deformation modes, which are stated in (d).

By comparing the SEM micrographs of both deformed samples, explicit differences in the deformation morphology are observed (see Fig. D.4(a) and (b)). While pure Au deforms very homogeneous, for the alloy a pronounced shear band pattern emerged during deformation. The pattern, with shear bands preferentially oriented under 45°-50° with respect to the loading direction, spreads over the entire film. Relating these observations to the in-plane evolution of the normalized integral (111) peak intensity

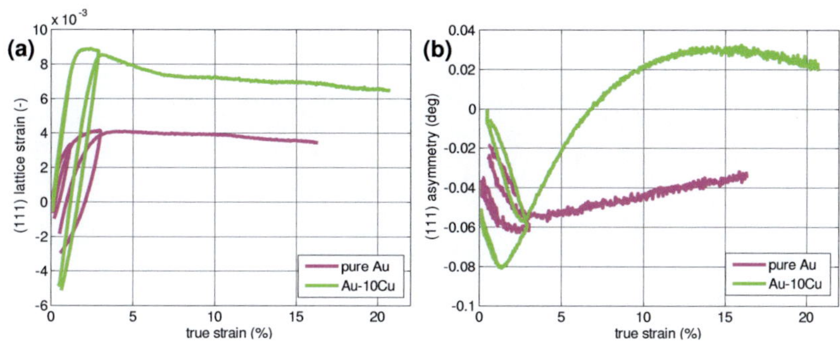

Figure D.3.: Results from *in situ* tensile testing of pure Au and Au-10Cu. (a) (111) lattice strain and (b) (111) peak asymmetry over ε for $\phi = 90°$. Pure Au shows distinctly lower lattice strain values, however, also the relaxation at high strains is much less pronounced. The behavior of peak asymmetry is likewise distinctly different: Pure Au does not show a pronounced reversal, as Au-10Cu as well as Pd and PdAu alloys do (cf. Fig. 5.18).

(shown in the polar plots of Fig. D.4(c) and (d)), it is found that the formation of in-plane deformation texture, observed for the alloyed sample, coincides with shear band formation. On the other hand, the homogeneous deforming Au film, only exhibits very slight intensity redistributions.

As it was discussed in section 5.3.1 for NC Pd thin films, the onset of texture formation coincides with the reversal of asymmetry, and both parameters are indicative for dislocation-based plasticity. Therefore, it is argued that for pure Au dislocation plasticity is not as active and is mostly replaced by other deformation mechanisms. One possibility are certainly GB-mediated deformation processes (shear and slip) or GB migration, since no solutes are present which could pin GBs. Moreover, the Au film is in the similar grain size range as the Pd films from chapter 5, however, the melting point of Au is considerably lower than for Pd, leading to an increased homologous temperature at RT of 0.22, while for Pd and Ni 0.16 and 0.17 is calculated, respectively. Therefore, one could speculate

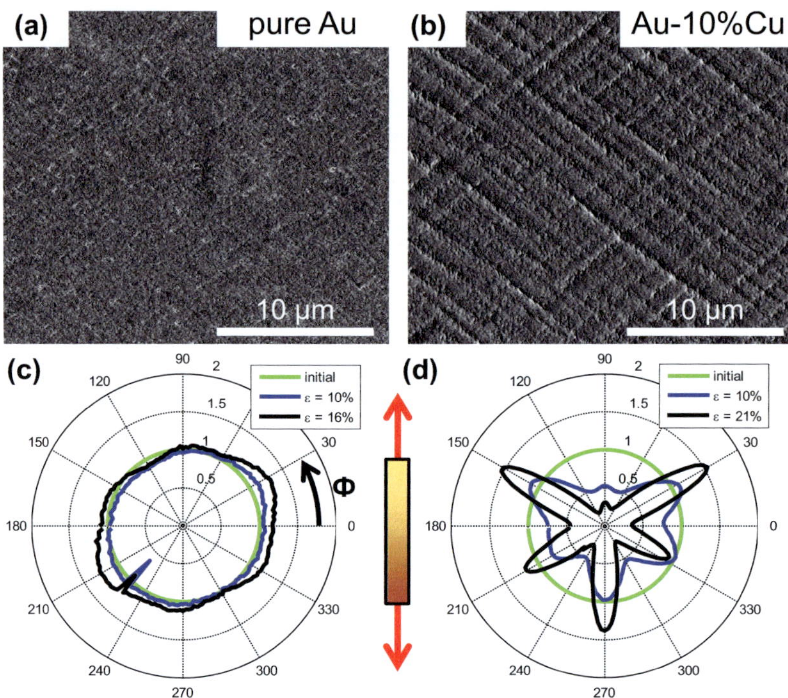

Figure D.4.: SEM micrographs of the deformed samples show (a) a homogeneously deformed pure Au film, and (b) a periodic and highly oriented shear band pattern spreading the entire sample for the alloyed sample. In (c) and (d) the evolution of the in-plane (111) integral peak intensity is displayed in polar plots for both samples. Obviously, shear band formation is related to formation of deformation texture.

on diffusion processes, which were also discussed in Ref. [Gruber et al., 2008b] for tensile tested pure Au in thin film geometry. Since film thickness of the herein tested samples are in the similar range, but grain size is considerably smaller ($D \approx 30\,\text{nm}$), the role of diffusion could be even more pronounced. In fact, the reversal of peak asymmetry is much weaker in the Au film (see Fig. D.3(b)) compared to Au-10Cu, emphasizing the reduced activity of dislocation-based plasticity.

The in-plane intensity variation observed by XRD, is proven by ACOM in conjunction with TEM (see Appendix A for methodological details). For this purpose, plane-view TEM lamellae were prepared by FIB. From orientation maps, shown in Fig. D.5(a) and (b), the orientation density of different (hkl) families in each individual in-plane direction can be extracted. The (111) in-plane orientation densities are displayed in D.5(c) for pure Au and in Fig. D.5(d) for the AuCu alloy, and compared to the normalized (111) integral XRD peak intensities. Comparing the deformed films, both methods yield a relatively uniform distribution for pure Au, while for Au-10Cu the six-fold symmetry is confirmed by both methods. However, differences in the sharpness of the six-fold symmetry are detected. One possible explanation could be that the recorded orientation map is only $1000\,\text{nm} \times 650\,\text{nm}$, which is more than a factor 10^5 smaller, than the area probed by XRD. The shear band width is around $1\,\mu\text{m}$, and the spacing between the bands is approximately of equal size. If the orientation map is incidentally recorded in a less sheared region, the resultant six-fold symmetry from azimuthal-dependent orientation densities will be less sharp. Nevertheless, both methods yield comparable results.

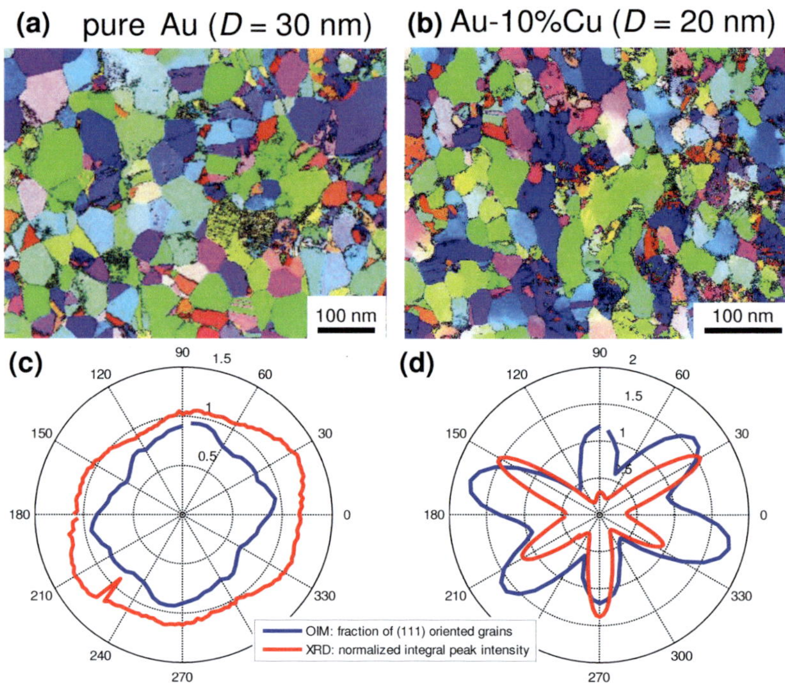

Figure D.5.: Results from ACOM/TEM investigations of the deformed samples. Orientation maps were recorded for (a) the pure Au film, and (b) the Au-10Cu film. In (c) and (d) the in-plane distribution of (111) oriented grains is compared to the in-plane (111) peak intensity distribution revealed by in situ XRD, both showing similar redistribution behavior.

In summary, Au and AuCu NC thin films were *in situ* tensile tested with synchrotron-based XRD techniques. From peak shape analysis and complementary microscopic investigations the following conclusions can be drawn.

- During deformation of the highly ductile thin films, different deformation modes can be separated to: (i) elastic behavior, (ii) shear band formation, and (iii) film cracking.

- The transition strains of the regimes are clearly alloy-dependent with an extended elastic and a reduced shear banding regime for higher Cu contents. Consequently, earlier film cracking is observed for those alloys. On the other hand, film cracking is not observed for the most ductile films in the investigated strain range ($\varepsilon_{max} = 30\%$).

- The shear banding mechanism involves the formation of an in-plane deformation texture indicating dislocation-based plasticity in AuCu alloys with $D \approx 20\,\text{nm}$.

- Surprisingly, for pure Au ($D \approx 20\,\text{nm}$) neither an obvious texture, nor pronounced shear bands, nor a distinct asymmetry reversal are observed; all indications for limited dislocation plasticity, although an aggravated geometrical confinement compared to AuCu is inexistent. Rather likely is the upcoming ascendence of a different deformation mechanism taking advantage of the large GB network and the pure microstructure.

E. Localized Deformation of Nanotwinned Cu

Two deformed nanotwinned (NT) bulk Cu samples were scanned with the microfocused X-ray beam (beam size: $8 \times 20\,\mu m^2$) available at the High Energy Microdiffration (HEMD) endstation of beamline ID15A at the ESRF (see section 3.2.1 for details on the setup). The as-deposited microstructure of the samples consists of highly aligned, stacked nanotwins, with a twin spacing of $\approx 40\,nm$ and an in-plane domain size of $\approx 500\text{-}800\,nm$. For details on the sample preparation and the initial microstructure, the reader is referred to Refs. [Hodge et al., 2006, 2008]. The samples were both tensile tested with strain rates of $\approx 10^{-4}\,s^{-1}$, one sample at liquid nitrogen (LN2) temperature (77 K) and the other at room temperature (RT) [Hodge et al., 2011]. The sample tested at LN2 exhibits a prolonged necked area and failed along a shear band, while the sample tested at RT deformed more homogeneous and fails by shear fracture under $45°$ with respect to the loading direction. Subsequent to mechanical testing, different XRD scans parallel and perpendicular to the loading direction were conducted with a step size of 10 - $20\,\mu m$. The experiments are embedded in a collaboration with the group of Prof. Dr. Andrea M. Hodge (University of Southern California) and FIB investigations were carried out by Timothy A. Furnish.

By scanning the sample, which was tensile tested at LN2 temperature, parallel to the gauge section and loading direction, the recorded diffraction patterns (DPs) change significantly. The deformation can be separated into three regions (see Fig. E.1(a)-(c)). Within a distance $\Delta d < 120\,\mu m$ from the edge of the fractured shearband, only individual spots appear as diffraction pattern, instead of complete Debye-Scherrer rings. For $120\,\mu m < \Delta d < 1550\,\mu m$, a pronounced texture is observed for all reflections. The texture consists of six-fold symmetries for each reflection, with maxima in intensity for (111), (200), (311), and (222) in tensile direction, while (220) shows minima in tensile direction. This is similar to a typical deformation texture for coarse-grained FCC metals under tensile load [Barrett and Massalski, 1966]. For $\Delta d > 1550\,\mu m$, the DPs are similar to DPs recorded in the undeformed region, exhibiting continuous rings, however with rather grainy than smooth contours.

The comparison of the diffraction patterns with the corresponding micrographs from FIB cross sections (Fig. E.1(a2)-(c2)) taken at the same distances from the fracture site allows to identify different deformation modes: In the direct vicinity of the shear band, the twinned structure is completely dissolved and micrometer-sized grains are present. In the prolonged, necked area the pronounced deformation texture indicates considerable dislocation activity. By inspection of the corresponding micrograph (Fig. E.1(b2)), it is reasoned that the incipient detwinning process is carried by dislocation processes, as the highly aligned undeformed microstructure (Fig. E.1(c2)) is free of any in-plane texture (Fig. E.1(c1)).

On the other hand, from the scans of the sample tested at RT only two different regions are identified, see Fig E.2. No region with individual diffracting spots and large grains is apparent. Instead, for $\Delta d < 500\,\mu m$, the same texture is observed, as for the LN2 sample in the necked area. For $\Delta d > 500\,\mu m$, the DPs are again similar to the undeformed region.

In summary, by scanning deformed NT Cu samples with a microfocused X-ray beam, the deformation behavior can be resolved locally and different deformation zones can be identified. Comparison of two samples, tested at LN2 and room temperature, respectively, reveals that the detwinning

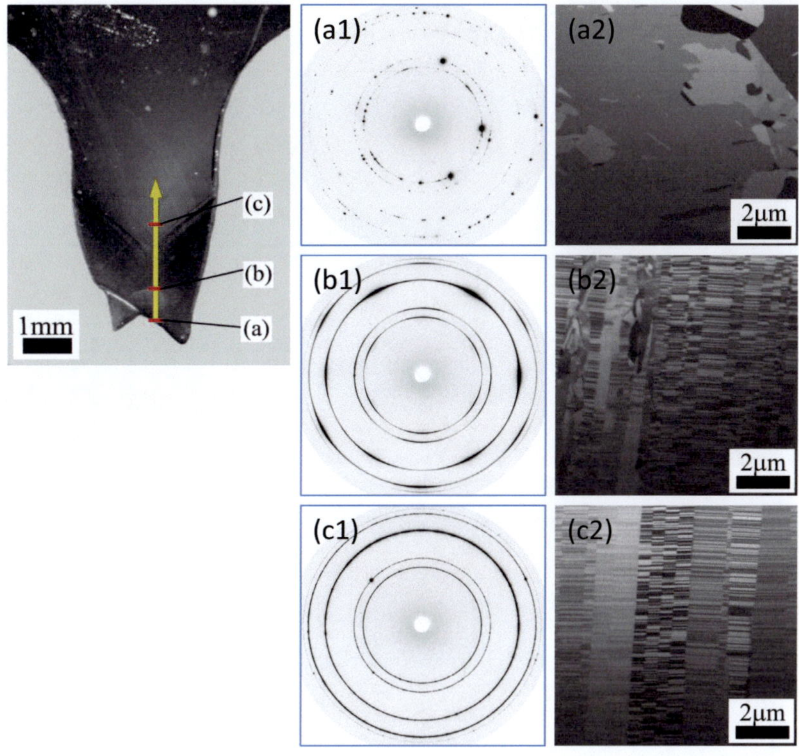

Figure E.1.: Nanotwinned bulk Cu sample tensile tested at liquid nitrogen temperature. Scanning using a microfocused X-ray beam reveals different deformation regimes, indicated by (a1)-(c1) three representative diffraction patterns taken at a distance from the fracture edge of $\Delta d = 50\,\mu m, 700\,\mu m$, and $2\,mm$, respectively. (a2)-(c2) Corresponding micrographs from FIB cross sections are presented for comparison. In the vicinity of the shear band, the initially highly twinned microstructure is completely dissolved and micrometer sized grains are present. In the necked area the strong deformation texture indicates that dislocations are responsible for the incipient detwinning process. Outside the necked region, the microstructure, and respectively the DP, are similar to the undeformed counterpart.

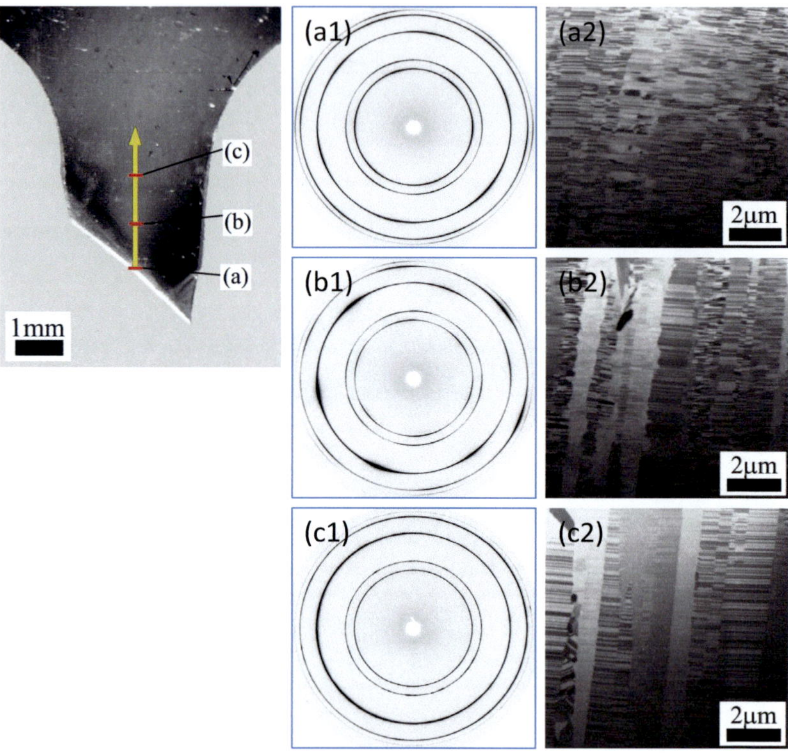

Figure E.2.: Nanotwinned bulk Cu sample tensile tested at room temperature. (a1)-(c1) The diffraction patterns and (a2)-(c2) corresponding micrographs are recorded at a distance $\Delta d = 50\,\mu m$, $150\,\mu m$, and $1\,mm$, respectively. No fully detwinned area can be identified, and the partially detwinned area (necked area) is much more localized to the shear-affected region close to the fracture site, compared to the sample tested at LN2.

processes seems to be benefited by low testing temperatures, as detwinning is much more advanced in the sample tested at 77 K. In the LN2 sample, a completely detwinned area in the direct vicinity of the shear band ($\Delta d < 120\,\mu m$) and a prolonged necked area, where dislocation activity dissolves the twinned structure were identified. In contrast, in the RT sample no completely detwinned zone is identified and moreover the partial detwinning is not that advanced but rather localized to the fracture site.

List of Figures

List of Tables

List of Symbols and Abbreviations

β .... Geometric Coupling Factor

β_G .... Gaussian Share of Integral Breadth

β_L .... Lorentzian Share of Integral Breadth

ΔA .... Absolute Change of Asymmetry

δ .... Grain Boundary Thickness

$\dot{\varepsilon}$ .... Strain Rate

ε .... True Strain

ε_c .... Total Compressive Strain

ε_{hkl} .... Elastic Lattice Strain

γ_{sf} .... (Stable) Stacking Fault Energy

γ_{usf} .... Unstable Stacking Fault Energy

λ .... X-Ray Wavelength

ν .... Poisson's Ratio

ν_{GB} .... GB Volume Fraction

ϕ .... Azimuthal, In-Plane Angle

ρ .... Dislocation Density

σ .... Stress

σ_0 .... Bulk Strength

σ_y .... Yield Strength

θ .... Bragg's Angle

A .... Peak Asymmetry

A_0	....	Peak Amplitude
A_{GB}	....	GB Area
D	....	Grain Size
d	....	Characteristic Length
d_{hkl}	....	(hkl) Lattice Spacing
E	....	Young's Modulus
E_{Xray}	....	Energy of the X-ray beam
I	....	Intensity / Counts
V_{GB}	....	GB Volume
$<\varepsilon>$	....	Root Mean Square microstrain
ACOM	....	Automated Crystal Orientation and phase Mapping
ANKA	....	ANgströmquelle KArlsruhe
BCC	....	Body Centered Cubic
CCD	....	Charge-Coupled Device
CG	....	Coarse-Grained
CMOS	....	Complementary Metal-Oxide-Semiconductor
COMP	....	COMPression specimen (Ni reference sample for SCS)
DC	....	Direct Current
DDD	....	Discrete Dislocation Dynamics
DFG	....	German Research Foundation
DIC	....	Digital Image Correlation
DP	....	Diffraction Pattern
EDX	....	Energy-Dispersive X-ray spectroscopy
ESRF	....	European Synchrotron Radiation Facility
FCC	....	Face Centered Cubic
FEM	....	Finite Element Method
FIB	....	Focused Ion Beam

FWHM	Full Width at Half Maximum
GB	Grain Boundary
GBS	Grain Boundary Sliding
HCP	Hexagonal Close Packed
HEMD	High Energy MicroDiffraction
HP	Hall Petch
HPT	High Pressure Torsion
IBR	Integral peak BRoadening
IGC	Inert Gas Condensation
INT	Integral peak INtensity
MD	Molecular Dynamic
NC	NanoCrystalline
NT	NanoTwinned
PED	Pulsed Electrodeposition
RF	Radio Frequency
SCS	Shear-Compression Specimen
SEM	Scanning Electron Microscopy
SFE	Stacking Fault Energy
SLM	Single Line Method
SLS	Swiss Light Source
TB	Twin Boundary
TEM	Transmission Electron Microscopy
UFG	Ultra Fine Grained
WA	Warren-Averbach
WH	Williamson-Hall
XRD	X-Ray Diffraction

Schriftenreihe
des Instituts für Angewandte Materialien

ISSN 2192-9963

Die Bände sind unter www.ksp.kit.edu als PDF frei verfügbar oder
als Druckausgabe bestellbar.

Band 1   Prachai Norajitra
Divertor Development for a Future Fusion Power Plant. 2011
ISBN 978-3-86644-738-7

Band 2   Jürgen Prokop
**Entwicklung von Spritzgießsonderverfahren zur Herstellung von
Mikrobauteilen durch galvanische Replikation.** 2011
ISBN 978-3-86644-755-4

Band 3   Theo Fett
**New contributions to R-curves and bridging stresses – Applications
of weight functions.** 2012
ISBN 978-3-86644-836-0

Band 4   Jérôme Acker
**Einfluss des Alkali/Niob-Verhältnisses und der Kupferdotierung auf
das Sinterverhalten, die Strukturbildung und die Mikrostruktur von
bleifreier Piezokeramik $(K_{0,5}Na_{0,5})NbO_3$.** 2012
ISBN 978-3-86644-867-4

Band 5   Holger Schwaab
**Nichtlineare Modellierung von Ferroelektrika unter Berücksich-
tigung der elektrischen Leitfähigkeit.** 2012
ISBN 978-3-86644-869-8

Band 6   Christian Dethloff
**Modeling of Helium Bubble Nucleation and Growth in Neutron
Irradiated RAFM Steels.** 2012
ISBN 978-3-86644-901-5

Band 7   Jens Reiser
**Duktilisierung von Wolfram. Synthese, Analyse und Charakterisierung
von Wolframlaminaten aus Wolframfolie.** 2012
ISBN 978-3-86644-902-2

Band 8   Andreas Sedlmayr
Experimental Investigations of Deformation Pathways in Nanowires.
2012
ISBN 978-3-86644-905-3

Band 9   Matthias Friedrich Funk
           **Microstructural stability of nanostructured fcc metals during cyclic
           deformation and fatigue.** 2012
           ISBN 978-3-86644-918-3

Band 10   Maximilian Schwenk
           **Entwicklung und Validierung eines numerischen Simulationsmodells
           zur Beschreibung der induktiven Ein- und Zweifrequenzrandschicht-
           härtung am Beispiel von vergütetem 42CrMo4.** 2012
           ISBN 978-3-86644-929-9

Band 11   Matthias Merzkirch
           **Verformungs- und Schädigungsverhalten der verbundstranggepressten,
           federstahldrahtverstärkten Aluminiumlegierung EN AW-6082.** 2012
           ISBN 978-3-86644-933-6

Band 12   Thilo Hammers
           **Wärmebehandlung und Recken von verbundstranggepressten
           Luftfahrtprofilen.** 2013
           ISBN 978-3-86644-947-3

Band 13   Jochen Lohmiller
           **Investigation of deformation mechanisms in nanocrystalline
           metals and alloys by in situ synchrotron X-ray diffraction.** 2013
           ISBN 978-3-86644-962-6